# Water and Wastewater Treatment

# Water and Wastewater Treatment

Editors

**Krzysztof Barbusiński**
**Maciej Thomas**
**Krzysztof Filipek**

MDPI • Basel • Beijing • Wuhan • Barcelona • Belgrade • Manchester • Tokyo • Cluj • Tianjin

*Editors*

Krzysztof Barbusiński
Silesian University of Technology
Poland

Maciej Thomas
Cracow University of Technology
Poland

Krzysztof Filipek
Krzysztof Filipek Advisory Company
Poland

*Editorial Office*
MDPI
St. Alban-Anlage 66
4052 Basel, Switzerland

This is a reprint of articles from the Special Issue published online in the open access journal *Clean Technologies* (ISSN 2571-8797) (available at: https://www.mdpi.com/journal/cleantechnol/special_issues/water_wastewater_treatment).

For citation purposes, cite each article independently as indicated on the article page online and as indicated below:

LastName, A.A.; LastName, B.B.; LastName, C.C. Article Title. *Journal Name* **Year**, *Volume Number*, Page Range.

**ISBN 978-3-0365-3481-7 (Hbk)**
**ISBN 978-3-0365-3482-4 (PDF)**

# Contents

# About the Editors

**Krzysztof Barbusiński** is a Full Professor at the Department of Water and Wastewater Engineering (Silesian University of Technology; Poland), where he conducts research and lectures in the field of municipal and industrial wastewater treatment. His scientific achievements include over 300 publications in scientific journals and proceedings, 3 patents, and many research projects and industrial implementations. Moreover, he has many years of experience as an expert and scientific consultant in cooperation with enterprises and research institutes. His main research interests includeinnovative wastewater and landfill leachate treatment methods, advanced oxidation processes, the removal of nitrogen (in deammonification process) released from digested sewage sludge, the application of the circular economy in WWTPs, and the recovery of water and raw materials from wastewater and sewage sludge. At present, he is a member of the Sanitary Engineering Section (Committee of Civil and Water Engineering) of the Polish Academy of Sciences.

**Maciej Thomas**, Ph.D., is a researcher at the Department of Environmental Technologies, Faculty of Environmental Engineering and Energy, Cracow University of Technology, Cracow, Poland. He is a specialist in the field of industrial water and wastewater treatment and has several years of practical experience. His scientific achievements include over 50 scientific papers and numerous expert assessments and industrial implementations. His research involves the use of green chemical oxidants for the treatment of water and wastewater that has been contaminated with heavy metals and toxic organic compounds. His research is presently focused on the synthesis and use of potassium ferrate to oxidize hazardous contaminants in industrial wastewater. Additionally, Dr. Thomas' research involves the synthesis and application of trithiocarbonic acid derivatives for thhe precipitation of heavy metal ions from wastewater originating from the galvanic and printed circuit boards industry.

**Krzysztof Filipek**, Ph.D., is an expert in the field of water and wastewater economy as well as in water and wastewater treatment, rainwater management, water safety plans, and risk analysis in water distribution systems. He is an author and co-author of about 50 papers and has a great of expertise in the field of water and wastewater management. His main research interests comprise municipal wastewater treatment, ground water treatment, and sludge processing, and he also provides legal solutions to environmental disputes. Currently, he is an advisor to water companies, municipalities, and industrial entities. Moreover, he is a post-graduate studies lecturuer at the University of Silesia and WSB University, Poland.

*clean technologies*

MDPI

*Editorial*

# Water and Wastewater Treatment: Selected Topics

**Krzysztof Barbusiński [1,*] and Krzysztof Filipek [2]**

1 Department of Water and Wastewater Engineering, Silesian University of Technology, Konarskiego 18, 44-100 Gliwice, Poland

2 Advisory Company, Wróblewskiego 8/27, 41-106 Siemianowice Slaskie, Poland; k.filipek@fdkf.pl

* Correspondence: krzysztof.barbusinski@polsl.pl

**Citation:** Barbusiński, K.; Filipek, K. Water and Wastewater Treatment: Selected Topics. *Clean Technol.* **2022**, *4*, 91–96. https://doi.org/10.3390/cleantechnol4010007

Received: 29 January 2022
Accepted: 8 February 2022
Published: 14 February 2022



Currently, there is a real need for rapid progress and development in almost all industries and areas of human activity. In the field of wastewater treatment, the development of already existing technologies is clearly recognized. The main focus is on the development of new, innovative processes concerning not only the degradation of pollutants, but also the recovery of valuable raw materials from wastewater. It is strongly linked to carbon footprint reduction and it also refers to the recently discussed "water footprint".

Until recently, only "wasted water" was considered as wastewater, and the purpose of purification was to remove organic contaminants, as well as nitrogen and phosphorus substances, and additionally, from industrial wastewater, specific impurities. Nowadays, wastewater and sewage sludge are more frequently being recognized as sources of energy and valuable, most often being non-renewable resources. Such an approach is consistent with the assumptions of circular economy ideas that could be applied in wastewater treatment plants by means of energy production and its optimized consumption and the recovery of important raw materials, e.g., phosphorous, nitrogen, and water from wastewater and sludge. In the near future, it could also be possible to recover bioplastic materials (polyhydroxyalkanoates; PHAs) and produce "green" hydrogen (the fuel of the future). Therefore, nowadays, modern wastewater treatment plants should not be considered only as facilities protecting water ecosystems, but also as technological energy plants and objects to recover important raw materials.

Moreover, there are some new challenges to be met by the wastewater treatment plants, e.g., the removal of pharmaceuticals. Currently, the purification processes carried out in wastewater treatment plants are not adapted to remove impurities such as, e.g., endocrine-disrupting compounds (EDCs) and pharmaceutical and personal care products (PPCPs). Therefore, in the near future, it is necessary to modernize the technological systems of treatment plants and to use more effective treatment techniques, such as advanced oxidation processes (AOPs), membrane processes, adsorption, etc.

It is also important to introduce the above-mentioned advanced methods to facilitate the effective treatment of industrial wastewater and landfill leachate. The intensification of activities leading to the use of treated industrial wastewater for re-use as process water is particularly important.

The presented Special Issue includes publications to cover a comprehensive range of wastewater treatment technologies. It contains three scientific publications based on the obtained research results and two review publications. The published articles cover a wide range of topics, which confirms that the wastewater treatment technology has significant interdisciplinarity and it is a field where there is significant progress in the research and implementation of innovative scientific and technical solutions.

We provide a brief review of the papers published as follows:

*Waste Ochre for Control of Phosphates and Sulfides in Digesters at Wastewater Treatment Plants with Enhanced Biological Phosphorus Removal*

Öfverström et al. [1] used waste ochre to optimize anaerobic sludge digestion. They demonstrated the potential for using waste ochre instead of commercially available iron to reduce the concentration of hydrogen sulfide in biogas production and to reduce the release of phosphate into a sludge liquor at WWTPs with enhanced biological phosphorus removal. Ochre ($Fe_2O_3$) is a waste product from water treatment plants, based on iron-contaminated raw groundwater treatment. The authors used the ochre from the Antaviliai water treatment plant (Vilnius, Lithuania), which supplies drinking water to 250,000 people. The ochre had a total solids content (TS) of about 40%, and 1 g TS of ochre contained approximately 350–400 mg $Fe^{3+}$. Batch and continuous pilot-scale tests were performed for the mesophilic digestion of primary and waste-activated sludge with different doses of ochre.

It has been shown that in order to prevent the inhibition of the methane production process, small doses of ochre should be applied continuously, and the dosing should be optimized in order to reduce extra sludge production. In batch tests, the dosing of ochre (0.5 g $Fe^{3+}$/L and 1 g $Fe^{3+}$/L) showed no inhibition of biogas production. With the addition of ochre, the accumulated methane potentials in the reactors were 10–15% higher than the methane potential in the control reactor. During the batch test, the phosphate release was reduced by 29% and 57% for the low and high doses, respectively.

During the pilot scale experiment, an immediate drop from 2000 ppm down to 570 ppm of the $H_2S$ concentration in the biogas was seen after the addition of ochre at two different doses: 2.5 g $Fe^{3+}$/d and 5 g $Fe^{3+}$/d. However, the anaerobic conversion process in the reactor with the highest dose (5 g $Fe^{3+}$/d) was inhibited by the ochre, resulting in high acetate concentrations (230–1700 mg/L). In a second pilot scale experiment, ochre was dosed continuously in smaller amounts (1.5 and 0.75 g $Fe^{3+}$/d) to avoid any inhibition processes during the phosphate precipitation. A reduction in phosphates in the sludge liquor (33% and 66% for the low and high doses, respectively) was seen.

Hence, the described results show the technological phenomena related to the effective fermentation of sewage sludge and biogas production, as well as prevented the release of phosphorus from the sludge into post-fermentation leachate. The increased amount of phosphorus returned to the main technological line with these leachates is a clear signal for the operators of wastewater treatment plants to increase the doses of coagulants for chemical phosphorus precipitation from wastewater in order to meet the quality conditions of the discharged wastewater. In turn, hydrogen sulfide should be effectively removed from biogas because it is known to destroy the metal parts of cogeneration units used for the integrated production of heat and electricity in a sewage treatment plant. What is particularly worth noting is the use of a waste material (waste ocher), which is an example of introducing a circular economy in wastewater treatment plants.

*Lomefloxacin—Occurrence in the German River Erft, Its Photo-Induced Elimination, and Assessment of Ecotoxicity Subsection*

One of the modern tasks of wastewater treatment technology is the implementation of effective methods for the removal of pharmaceuticals and their metabolites in wastewater treatment plants. The technologies currently used in wastewater treatment plants are not adapted to the removal of this type of micro-pollutant. The processes of surface water treatment are not dedicated to removing these micro-pollutants as well. Thus, the pharmaceuticals and their metabolites could easily be transported into the natural water cycle, and in the form of drinking water could enter into human and animal organisms, which may cause a number of health problems [2]. Chemically, pharmaceuticals and their metabolites belong to various chemical groups and no uniform method of their removal or utilization has been developed so far. Currently, the Advanced Oxidation Processes (AOPs) are being used for this purpose, but it is most often carried out at the laboratory scale, because there are no legal regulations to oblige operators of the wastewater treatment plant to implement effective methods of pharmaceuticals removal. In addition, all the known methods do not remove individual pharmaceuticals with a similar efficiency [3].

An alternative to AOP processes can be, e.g., the use of potassium ferrate (VI) ($K_2FeO_4$) as a strong oxidant [4].

Lomefloxacine is an antibiotic from the fluorchinolone group. These compounds, once they have reached the water reservoirs, are only scarcely decomposed by living microorganisms, and are able to survive for a long time to interact with the other water organisms [5]. The main threat of this situation refers to its continuous acting. Despite the low concentrations, at the level of ng/L, the long lasting supply with the discharged wastewater may cause real problems—the so-called pseudo-persistent compounds. The results of such a long-term exposure, however, at relatively low doses, almost non-observable and detected at very low concentrations only, could be statistically important and detected for the next many generations of the ecologically important organisms. It could also lead to constant and irreversible changes extending the adaptability of many animal species present in the streams, rivers and lakes.

Voigt et al. [6] in their publication researched a very interesting and current topic concerning the degradation of the antibiotic Lomefloxacin (LOM), detected in the German river Erft. The methodology of near and far ultraviolet (UVA, UVC) radiation was used as AOPs and examined in relation to pH, water matrix, and catalysts. AOP catalysts, hydrogen peroxide and titanium dioxide were used. A chemical kinetics description revealed that UVC at pH 8–9 led to the fastest degradation of LOM. The catalysts hydrogen peroxide and titanium dioxide only had a limited influence on the degradation rate. Seven novel transformation products of the examined reactions were structurally identified by high-resolution higher-order mass spectrometry. The ecotoxicity of the novel and known compounds was assessed by quantitative structure–activity relationship (QSAR) analysis. In addition, the irradiation time-dependent minimum, and half-maximum inhibitory concentrations (MIC, IC50) of LOM solutions were clearly determined and suggested as ecotoxicological hazard indicators. From MIC and kinetic rate constants, the irradiation time required for compound and activity removal could be predicted.

In the Guest Editors' opinion, the reviewed paper should be distinguished as a crucial one in terms of water ecosystem protection from antibiotics. On the other hand, it is another reason to search the advanced methods to remove antibiotics from the wastewater, as was pointed out in the preface of this review.

*Stimulating Nitrogen Biokinetics with the Addition of Hydrogen Peroxide to Secondary Effluent Biofiltration*

Friedman et al. [7] used a tertiary wastewater treatment to remove $NH_4^+$, $NO_2^-$ and organic matter from secondary effluents in a pilot bio-filtration system assisted by adding hydrogen peroxide ($H_2O_2$). The authors examined the impact of adding $H_2O_2$ as a supplemental oxygen source on the operational efficiency of a secondary effluent filtration system with biologically active media (biofiltration). The study examined the feasibility of stimulating microbial activity using $H_2O_2$ as a bio-specific and clean oxygen source that leaves no residuals in water and is advantageous upon aeration due to the solubility limitations of oxygen. The tertiary wastewater treatment system consisted of the following elements: wire filter (500 µm mesh), the coagulator–flocculator tank (with addition of polyaluminum chloride) and the biofilter bed with short HRT. The performance of a pilot bio-filtration system at a filtration velocity of 5–6 m/h was enhanced by the addition of $H_2O_2$ for particle, organic matter, $NH_4^+$ and $NO_2^-$ removal. Hydrogen peroxide provided the oxygen demand for full nitrification. As a result, influent concentrations of 4.2 mg/L N-$NH_4^+$ and 0.65 mg/L N-$NO_2^-$ were removed in the biofilter. The biofiltration without $H_2O_2$ addition only removed up to 0.6 mg/L N-$NH_4^+$ and almost no N-$NO_2^-$. Thus, the system showed a significant removal of $NO_2^-$ and $NH_4^+$ when $H_2O_2$ was added in comparison to the control system (without $H_2O_2$). The authors also presented a model to describe the biokinetics of tertiary wastewater biofiltration systems with the addition of $H_2O_2$.

The research results presented in this publication could serve as a very good example of how to introduce a circular economy idea in the field of wastewater treatment technology.

This Special Issue contains two review articles.

*Removal of Odors (Mainly $H_2S$ and $NH_3$) Using Biological Treatment Methods*

The air contamination by the odors and the volatile organic compounds (VOC) has been considered as a very current and serious problem all over the world. The odor nuisance and the health hazards are the main topics concerning both specialists and ordinary people. The VOC and odor compounds are the most prominent group in the atmosphere contamination factors. These compounds are known to be the precursors of photochemical oxidation, responsible for the formation of tropospheric ozone and smog, as a final result. They are counted as cancerogenic substances. The VOC and odors are formed as the by-products of the numerous industrial processes in, e.g., the varnish, wood, chemical, pharmaceutical, petrochemical and paper industries. They are also present in wastewater treatment plants and in the solid wastes processing plants. The emission of these compounds is relatively lower than nitrogen oxides (NOx); they show, however, a higher toxicity and reactivity, making them real threats to the environment and all human beings.

Facilities emitting the most persistent harmful gases belong to the municipal sector, including wastewater treatment plants, waste management plants or composting plants. The odor-producing compounds include:

- Sulfur compounds, i.e., hydrogen sulfide, thiols, sulfides, and alkyl disulfides;
- Nitrogen compounds, i.e., ammonia, and aliphatic amines;
- Organic compounds, including aldehydes, ketones, and fatty acids (phenol, cresol, butyric acid, acetic acid, and valeric acid).

The more and more active public awareness and strict formal environmental rules render the effective and innovative methods to eliminate the air contamination components highly awaited.

Barbusiński et al. [8] reviewed the available and most commonly used methods of gas deodorization. Comparing various, physical, chemical and biological, methods of odor removal, biological methods of pollution degradation undoubtedly have a clear advantage over the others—chemical and physical methods. This advantage is manifested mainly in ecological and economic terms. The possibility of using biological methods to remove $H_2S$ and $NH_3$, as the most common emissions by the municipal sector companies, was analyzed in terms of their removal efficiency. The method of bio-purification of air in biotrickling filters is more advantageous than the others, due to the high effectiveness of VOCs and odors degradation, lack of secondary pollutants, and economic aspects; it is a method that competes with the commonly used air purification method in biofilters.

A survey of the literature revealed that the biological methods of odor removal lead to high rates of the bio-purification of air, up to 95–99%; moreover, their advantage is apparent innovation, primarily in the economic aspect as well as in terms of environmental friendliness. Therefore, it is necessary to develop and intensify processes based on the biological methods of odor removal, in order to implement them on the full industrial scale. A good example of the modern biotechnologies methods for odor removal is the Compact Trickle Bed Bioreactors with their potential versatility.

*Review of Methods for Assessing the Impact of WWTPs on the Natural Environment*

Bąk et al. [9] reviewed methods of assessing the impact of wastewater treatment plants on the environment. This paper discusses the possible impact of WWTPs on the environment. Among other problems, such issues as energy consumption, noise and the formation of bioaerosols and odor nuisances were taken into account. Different ways of assessing the impact of wastewater treatment plants on the environment were described, taking into account the need to assess not only the technological process itself but also the building constructions in the course of operation. The results of various methods of the environmental impact assessment of wastewater treatment plants in selected countries were also compared.

Any wastewater treatment plant exerts a certain environmental impact during the construction and operation stages. Therefore, there is a need to assess this impact not only at the design and construction phase, but also during the facility's operation. While such assessments are frequent at the investment planning stage, they tend to be neglected in the operational phase. On the other hand, control activities are carried out in the context of compliance with certain regulations concerning, for example, gas emissions.

Environmental management as a part of the operating management in wastewater treatment plants (WWTPs) could be a basis for implementing the *Deming cycle* (a concept of quality management consisting of continuous improvement taking place in four successive stages: planning, execution, checking, and improvement), and thus the constant improvement of the mitigation of the environmental impact. The correct diagnosis of the current operating state of the WWTPs, the identification of aspects that may have a measurable impact on the environment, and the impact assessment are key factors. The suggested direction of development to assess the environmental impact of wastewater treatment plants is the improvement of the LCA technique. It could be supported by the implementation of environmental management systems, with the supplementation of these actions through green building certification. Only a holistic approach to the issue will enable all the environmental aspects to be taken into account and thus contribute to maximizing the subsequent environmental benefits. It is also advisable, if possible, to undertake necessary steps to develop and implement a unified method of assessing the impact of sewage treatment plants on the environment. Parallel to this, some activities that encourage and stimulate the uptake of such challenges should be introduced. As a consequence, the processes carried out in wastewater treatment plants will become even more "clean", and the treatment plants themselves will become more environmentally friendly. Hence, the introduction of circular economy solutions in wastewater treatment plants is also of great importance.

Despite the relatively few articles in this Special Issue, it should be noted that these publications cover a very wide range of relevant issues and challenges in the field of wastewater treatment technology, as was pointed out earlier. This proves the important role this Special Edition plays in understanding the directions of further development of wastewater treatment technology.

**Author Contributions:** K.B. and K.F. contributed equally to this work. All authors have read and agreed to the published version of the manuscript.

**Funding:** This research received no external funding.

**Institutional Review Board Statement:** Not applicable.

**Informed Consent Statement:** Not applicable.

**Acknowledgments:** The task of editing articles for this Special Issue was a real pleasure and rewarding for us. We would also like to thank the Clean Technologies team and the reviewers for their technical and substantive effort and contribution.

**Conflicts of Interest:** The authors declare no conflict of interest.

## References

1. Öfverström, S.; Davidsson, Å.; Haghighatafshar, S.; Kjerstadius, H.; la Cour Jansen, J. Waste Ochre for Control of Phosphates and Sulfides in Digesters at Wastewater Treatment Plants with Enhanced Biological Phosphorus Removal. *Clean Technol.* **2020**, *2*, 8. [CrossRef]
2. Huber, M.M.; Canonica, S.; Park, G.; von Gunten, U. Oxidation of pharmaceuticals during ozonation and advanced oxidation processes. *Environ. Sci. Technol.* **2003**, *37*, 1016–1024. [CrossRef] [PubMed]
3. Sharma, V.K.; Mishra, S.M.; Ray, A.K. Kinetic assessment of the potassium ferrate (VI) oxidation of antibacterial drug sulfamethoxazole. *Chemosphere* **2006**, *62*, 128–134. [CrossRef] [PubMed]
4. Kliś, S.; Barbusiński, K.; Thomas, M.; Mochnacka, A. Application of Potassium Ferrate (VI) in the Treatment of Selected Water and Wastewater Pollutants—Short Review. *Archit. Civil. Eng. Environ.* **2020**, *1*, 129–138. [CrossRef]

5. Bielińska, M.; Nałęcz-Jawecki, G. Pharmaceuticals in the Environment. I: Evaluation of Toxicity of Three Fluoroquinolones to Duckweed Lemna Minor. *Biul. Wydz. Farm. WUM* **2009**, *4*, 24–30. Available online: http://biuletynfarmacji.wum.edu.pl/0904 Jawecki/Jawecki.pdf (accessed on 24 January 2022). (In Polish)
6. Voigt, M.; Hentschel, B.; Theiss, N.; Savelsberg, C.; Bartels, I.; Nickisch-Hartfiel, A.; Jaeger, M. Lomefloxacin—Occurrence in the German River Erft, Its Photo-Induced Elimination, and Assessment of Ecotoxicity. *Clean Technol.* **2020**, *2*, 6. [CrossRef]
7. Friedman, L.; Mamane, H.; Chandran, K.; Jekel, M.; Cikurel, H.; Hübner, U.; Elgart, M.; Dagan, S.; Santo-Domingo, J.; Avisar, D. Stimulating Nitrogen Biokinetics with the Addition of Hydrogen Peroxide to Secondary Effluent Biofiltration. *Clean Technol.* **2020**, *2*, 5. [CrossRef]
8. Barbusiński, K.; Parzentna-Gabor, A.; Kasperczyk, A. Removal of Odors (Mainly $H_2S$ and $NH_3$) Using Biological Treatment Methods. *Clean Technol.* **2021**, *3*, 9. [CrossRef]
9. Bąk, J.; Barbusiński, K.; Thomas, T. Review of Methods for Assessing the Impact of WWTPs on the Natural Environment. *Clean Technol.* **2021**, *3*, 7. [CrossRef]

*clean technologies*

MDPI

*Article*

# Waste Ochre for Control of Phosphates and Sulfides in Digesters at Wastewater Treatment Plants with Enhanced Biological Phosphorus Removal

**Svetlana Öfverström [1], Åsa Davidsson [2], Salar Haghighatafshar [2], Hamse Kjerstadius [3] and Jes la Cour Jansen [2,*]**

1 Department of Water Management, Vilnius Gediminas Technical University, Sauletekio av. 11, LT-10223 Vilnius, Lithuania; svetlana.ofverstrom@vgtu.lt

2 Water and Environmental Engineering, Department of Chemical Engineering, Lund University, Box 124, 22100 Lund, Sweden; asa.davidsson@chemeng.lth.se (Å.D.); Salar.Haghighatafshar@chemeng.lth.se (S.H.)

3 Nordvästra Skånes Vatten och Avlopp AB, Box 2022, 250 02 Helsingborg, Sweden; Hamse.Kjerstadius@nsva.se

* Correspondence: Jes.la_Cour_Jansen@vateknik.lth.se; Tel.: +45-45812806

Received: 5 January 2020; Accepted: 7 February 2020; Published: 6 March 2020

**Abstract:** Ochre, waste iron sludge from the treatment of iron rich groundwater for potable use, makes up a significant waste problem. Furthermore, wastewater treatment plants with enhanced biological phosphorus removal and the digestion of sludge are in lack of iron for the prevention of hydrogen sulfide production and the release of phosphorous during anaerobic digestion. Thus, the addition of ochre to anaerobic digestion is a potential beneficial reuse of ochre. Sludge from wastewater treatment plants with enhanced biological phosphorus removal was used for the experiments. Batch and continuous pilot-scale tests were performed for the mesophilic digestion of primary and waste-activated sludge with different doses of ochre. Two different doses of ochre corresponding to molar ratios of 1 and 2 moles $Fe^{3+}$/mole P released in the batch test resulted in 29% and 57% reductions of phosphates respectively in the sludge liquor compared to the control sludge without inhibiting the digestion process. In the pilot experiment, the dosing of ochre at both a high and low dose (molar ratios of 1.6 and 0.8 $Fe^{3+}/S^{2-}$, respectively) resulted in an immediate drop in the $H_2S$ concentration (from >2000 ppm down to 570 ppm), while the control reactor still produced biogas with a high hydrogen sulfide concentration. The inhibition of the digestion process was observed (accumulation of acetate) at the higher dose. In a second pilot scale experiment, lower doses of ochre were tested continuously (1.5 and 0.75 mole $Fe^{3+}$/mole $P_{released}$) to avoid any inhibition, while evaluating the phosphate precipitation. A reduction of phosphates in sludge liquor (33% and 66% for the low and high doses respectively) was obtained.

**Keywords:** waste ochre; biogas; enhanced phosphorus removal; hydrogen sulfide; phosphates precipitation

---

## 1. Introduction

In wastewater treatment plants (WWTPs) with enhanced biological phosphorus removal, the internal recycling of phosphates and potential production of hydrogen sulfide are limiting factors for the digestion of the surplus sludge. The internal recycling of sludge liquor with a high concentration of phosphate is usually the cause of the overloading of such plants with phosphates [1–4]. Phosphate concentrations of up to about 500 mg $PO_4$-P/L have been found in the sludge liquor due to the release of phosphates by phosphate accumulating organisms (PAOs) during anaerobic digestion [5], while normal concentrations are often below 10 mg $PO_4$-P/L in conventional WWTPs with the chemical precipitation of phosphorus. The anaerobic digestion of sludge from plants with biological phosphorus removal is also risky if sulfates are present in the feed sludge [6,7], since no chemicals are available to capture the sulfur in the digester. A high content of hydrogen sulfide in produced biogas can corrode the equipment and is a health risk for the personnel at the treatment plant [8,9].

The addition of commercial products based on iron chloride ($FeCl_3$) is the common preventive method for solving both problems [2,8–11]. In some reported cases, the addition of iron salts impaired the digestion process [12,13] and reduced the methanogens activity [14]. Other research [15,16] demonstrated small decreases in VS (volatile solids) degradation as well as a reduction in organic sulfur generation during the digestion process of sludge dosed with iron (Fe:P 3:1 molar ratio).

Ochre ($Fe_2O_3$) is a waste product from the production of potable water based on groundwater with a high content of reduced iron [17–19]. High concentrations of dissolved iron can cause an unpleasant taste, brown color of the water, and the corrosion of steel pipes and water-distribution networks [20]. Water with iron levels of more than 1 mg/L can cause health problems [21]. Typically, ochre is reduced to very low levels, and the iron rich sludge is normally disposed into landfills as waste [19,22]. A possible method for utilizing ochre is to use it as an alternative source of iron for the reduction of the hydrogen sulfide content, and for solving problems related to the internal recirculation of phosphates at WWTPs with biological phosphorus removal processes [19,23]. Since commercial precipitation chemicals based on iron or aluminum are expensive [19,24], ochre might be a good alternative in countries where biological phosphorus removal is required [25].

Many different chemical reactions might be involved when ochre is added to an anaerobic digester for the reduction of hydrogen sulfide and soluble phosphorous. A comprehensive presentation is given in [19], where ochre is assumed to dissolve in the presence of organic material under the reducing conditions in the digester. Furthermore, sulfide and phosphate are precipitated with the dissolved iron according to Equations (1) and (2):

$$Fe^{2+} + HS^{-} \rightarrow FeS + H^{+} \quad (1)$$

$$3\ Fe^{2+} + 2H_2PO^{-}4 \rightarrow Fe_3(PO_4)^2 + 4H^{+} \quad (2)$$

In Lithuania, potable water is largely based on iron rich groundwater [26]. The iron content in groundwater around Vilnius is about 1.1 mg/L, and the total phosphorus (total-P) in the influent to the WWTP is typically 5–10 mg/L, demonstrating that the supply of iron is insignificant on a molar basis, so that all ochre from the water supply can easily be utilized for wastewater treatment.

Demands on the effluent quality from wastewater treatment plants have been more stringent during recent years [2,4,27]. Furthermore, resource-saving and cost-reducing measures should be taken into account. One such measure could be to use ochre for the optimization of the anaerobic sludge digestion.

This study evaluates if waste ochre can be used to control hydrogen sulfide production, and to prevent the high internal recirculation of phosphorus at the growing numbers of treatment plants with biological phosphorus removal. Since the addition of ochre contributes to sludge production in the digesters and consequently reduces the retention time, there is a practical limit for the dose. Besides that limitation, the addition of ochre may lead to the inhibition of the methane production, just as other iron-based chemicals do [14–16,18,19]. The experiments were planned based on all these considerations. Initially, batch experiments were performed for the evaluation of any inhibitory effect and potential for the reduction of dissolved phosphorus through digestion. A subsequent experiment with a continuous operation in pilot scale was performed for the evaluation of the potential for rapid reduction in hydrogen sulfide formation, and we also conducted a longer experiment for the evaluation of the potential for reduction of the internal phosphorus recirculation. Together, the experiments shall demonstrate the potential use of ochre for solving significant problems for plants with biological phosphorus removal that digest the surplus sludge.

## 2. Experiment Section

Ochre from the Antaviliai water treatment plant (WTP) (Vilnius, Lithuania), which supplies drinking water for 250,000 people (~41% of Vilnius city inhabitants), was used in the experiments. The incoming groundwater contains about 1.1 mg/L of iron, which is reduced to 0.04 mg/L after treatment. The ochre used in this experiment had a total solids content (TS) of about 40%, and one g TS of ochre contained approximately 350–400 mg $Fe^{3+}$. Table 1 presents the major constituents of ochre from the analysis.

The sludge for the experiments was obtained from Öresundsverket WWTP in Helsingborg (Sweden). The plant treats wastewater from 120,000 people and from many different industries. The plant is operated with enhanced biological phosphorus removal and pre-denitrification, with hydrolyzed primary sludge as the additional carbon source [28]. No chemicals for phosphorus removal are used at the plant. A minor dose of $FeCl_3$ is added in the thickener to the primary sludge before digestion to prevent hydrogen sulfide ($H_2S$) formation in the digester. Primary sludge was therefore sampled before the addition of $FeCl_3$ and manually thickened, which resulted in a total solids (TS) content of 3.6% (average value), while waste-activated sludge had about 3% of TS. Sludge was mixed in the proportions 0.35:0.65 (volume basis) in accordance with the sludge production at the treatment plant. Fresh sludge was collected from the WWTP every week and stored in a cold room <+4 °C.

**Table 1.** The major constituents of ochre. In total, 48 elements were detected.

| Symbol | Element | Concentration (%) | Standard Deviation (%) |
|---|---|---|---|
| Fe | Iron [1] | 35.27 | 1.29 |
| Ca | Calcium | 4.63 | 0.12 |
| Si | Silicon | 2.53 | 0.16 |
| P | Phosphorus | 2.26 | 0.04 |
| Mn | Manganese | 0.38 | 0.01 |
| Al | Aluminum | 0.28 | 0.10 |
| S | Sulphur | 0.25 | 0.04 |
| Mg | Magnesium | 0.23 | 0.01 |
| Ti | Titanium | 0.14 | 0.01 |
| Ba | Barium | 0.12 | 0.00 |
| K | Potassium | 0.04 | 0.01 |
| Sr | Strontium | 0.03 | 0.00 |
| Cr | Chromium | 0.008 | 0.00 |
| Pb | Lead | 0.003 | 0.00 |
| Zn | Zinc | 0.003 | 0.00 |
| Cu | Copper | 0.002 | 0.00 |
| Ni | Nickel | 0.001 | 0.00 |
| As | Arsenic | 0.001 | 0.00 |
| Cd | Cadmium | B.D. | 0.00 |

[1] One g TS of ochre contains approximately 350–400 mg $Fe^{3+}$.

Inoculum used for the start-up of the pilot digesters came from the mesophilic reactors at Öresundsverket WWTP, operated at 35 °C. Inoculum used in the batch test was taken from the pilot-scale digesters after operation for more than three solid retention times (SRTs) in order to remove the effect of the iron used for sulfide control in the full-scale installation.

Total Solids (TS) and total volatile solids (VS) of the inoculum and sludge used for the batch and pilot experiments are presented in Table 2.

**Table 2.** The characteristics of the inoculum and the feed sludge used in the batch and pilot experiments.

| | Batch Experiment | | | Pilot Experiment | | |
|---|---|---|---|---|---|---|
| | Inoculum | Primary Sludge | Bio-Sludge | Inoculum | Primary Sludge | Bio-Sludge |
| $TS_{average}$ (%) | 1.65 | 4.84 | 3.84 | 2.75 | 3.58 | 3.27 |
| $VS_{average}$ (%) | 1.12 | 2.79 | 2.12 | 1.95 | 2.85 | 2.44 |

### 2.1. Digestion Tests

#### 2.1.1. Dose of Ochre

The dosing of ochre, as solids, in the batch experiments was set at two levels, Low (0.5 g $Fe^{3+}$/L) and High (1.0 g $Fe^{3+}$/L), where the lower dose corresponded to a molar ratio of 1:1 mole Fe/$P_{released}$. $P_{released}$ is the release of phosphate into the liquid phase during digestion, which was determined by the batch digestion of the sludge from Öresundsverket (around 260 mg $PO_4$-P/L). The high dose was chosen in order to evaluate any inhibitory effect and the effect of the increased dose on the phosphate binding capacity.

The dosing of ochre, as solids, during the pilot experiment was divided into two separate experiments. In the first period, high doses were added for only three days at two different levels in two different reactors, and a third was used as the control without any addition. The idea was to evaluate the potential to use ochre for the rapid reduction of hydrogen sulfide production and to check if very high doses of ochre have any impact on the anaerobic digestion process. In the second period, lower doses of ochre were added (two levels in two different reactors and one control without addition) during 15 days to evaluate the influence on the phosphate release.

#### 2.1.2. Laboratory Batch Test for Methane Potential

Batch tests were used for the examination of the methane potential according to the methodology described in [29,30]. Batch tests were performed in glass reactors of ~2 L. The tests were made under mesophilic conditions (35 °C). The total volume of inoculum, sludge, and ochre was about 500 mL, with the substrate representing about 45% of the total volatile solids (VS) content. The dose of ochre corresponded to 0.5 g $Fe^{3+}$/L and 1 g $Fe^{3+}$/L (1 and 2 mole $Fe^{3+}$ per mole of P released) of ochre, with the composition shown in Table 1. Each bottle was flushed with nitrogen gas after filling to ensure anaerobic conditions in the reactors. The duration of the experiment was 24 days. The production was followed by six measurements during the experiment, and each reactor was set in triplicates. Cellulose powder was used as the reference substrate to test the function of the inoculum. The amount of VS of cellulose added was similar to the VS of the tested substrate, according to the methodology described in [29,30].

#### 2.1.3. Pilot-Scale Digestion Experiment

Three pilot-scale reactors (20 L of working volume) were continuously operated at a mesophilic temperature (37 °C), solid retention time (SRT) of 15 d, and organic loading rate (OLR) of 1.42–2.00 kg VS/$m^3$·d, with separate biogas collection tanks for each reactor. Figure 1 shows the scheme of the pilot-scale reactors used for the experiment. The equipment was previously described in detail [31,32]. Each reactor was inoculated with anaerobic sludge from the mesophilic digesters at Öresundsverket. The reactors were fed once per day with 1.33 L of a mixture of primary and waste-activated sludge. In the first experiment concerning hydrogen sulfide reduction via addition of ochre, the doses were chosen at 5 g $Fe^{3+}$/d and 2.5 g $Fe^{3+}$/d in two of the reactors (corresponding to a molar ratio of 1.6 and 0.8 $Fe^{3+}/S^{2-}$, respectively). In the second period, involving the experiment for phosphates release control, the high and low doses were reduced to 1.5 g $Fe^{3+}$/d and 0.75 g $Fe^{3+}$/d respectively (corresponding to a molar ratio of 1.5 and 0.75 $Fe^{3+}/P_{released}$ respectively, where $P_{released}$ is 400 mg $PO_4$-P/L).

The pilot reactors were first operated continuously for 30 days to achieve steady state conditions and for the removal of the minor iron dose used at the full-scale installation. However, in order to avoid the transient phase in the results, the 30th day after the initial start-up was identified as day "1" in the data and the figures. The withdrawn sludge from the pilot scale reactors from day 26 was used as inoculum in the batch experiments. After a long period of operation, the hydrogen sulfide concentration in the biogas was still low, probably due to the low content of sulfates in the sludge feed. In order to provoke $H_2S$ formation, sodium sulfate ($Na_2SO_4$) was added daily in all three reactors, starting on day 28. The dosage of $Na_2SO_4$ in the reactors was increased gradually: 2 g $Na_2SO_4$/day was added for 5 days, 4 g $Na_2SO_4$/day for the next 10 days, and about 8 g $Na_2SO_4$/day was added until the end of the experiment (day 52). The hydrogen sulfide concentration in the produced biogas quickly increased after the $Na_2SO_4$ addition, and then the first experiment with the ochre dose for hydrogen sulfide control started and ran for 3 days. After a short rest period for the reactors (from days 53 to 63), the second part of the experiment started, and the reactors were operated without ochre dosing for 20 days (from days 64 to 84). Then, they were operated with relatively lower doses of ochre for the phosphates release control until the end of the experiment.

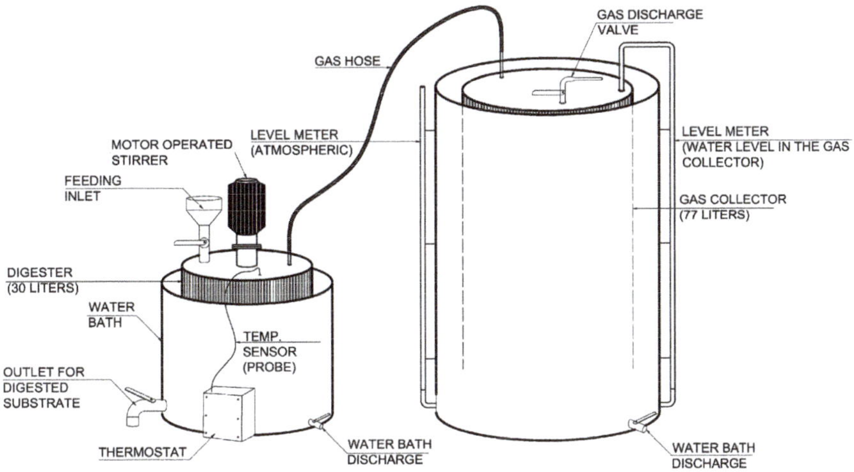


**Figure 1.** The schematic view of the pilot-scale equipment from [32]. Three identical reactors were used for the experiment.

## 2.2. Analytical Methods

The samples of primary activated and digested sludge from the batch and pilot-scale experiments were analyzed using HACH LANGE test tubes: LCK 320 for iron and LCK 049 for phosphate. The prepared tubes were analyzed with a HACH LANGE spectrophotometer (model DR 2800). All the samples stated above were centrifuged for 15 min at the speed of 10,000 rpm and filtrated through Munktell general purpose filter papers with a 6~10 µm pore size before further analysis. The TS and VS contents were measured according to standard methods. The alkalinity was measured according to the Swedish standard (ISO 9963-1:1994).

The methane production from the batch experiment was measured using a gas-tight syringe and gas chromatography (Varian 3800 Gas Chromatograph) equipped with TCD (thermal conductivity detector) and a column with dimensions of 2.0 m × 1/8 inch × 2.0 mm.

The composition of the produced biogas (methane, carbon dioxide, oxygen, and hydrogen sulfide (up to 2000 ppm)) during the pilot-scale experiment was measured using a portable gas-meter SEWERIN SR2-DO. The pH of the reactors was measured using a digital pH-meter (pH 3110 SET 2 incl. SenTiz® 41).

The volatile fatty acids (VFA) content of the samples was analyzed with gas chromatography using an Agilent 6850 Series GC System equipped with an FID (flame ionization detector) and a column with dimensions of 25 m × 0.32 µm × 0.5 µm.

## 3. Results and Discussion

### 3.1. Potential Use of Ochre

Ochre has a potential for reducing or preventing hydrogen sulfide formation in anaerobic digestion where iron deficit can enable such problems. Furthermore, the release of phosphorus during the anaerobic digestion of surplus sludge from plants with biological phosphorous removal can be prevented. However, it is necessary that the addition does not significantly influence the biogas production. Below, we present and discuss the experimental evidence for the potential.

### 3.2. Batch Experiment

The development of methane production in the batch experiments is shown in Figure 2 (the methane production from the inoculum has been extracted). The methane ($CH_4$) production with cellulose (reference substrate) was about 320 NmL $CH_4$/g $VS_{in}$ at the end of the experiment (after 24 days), which shows that the inoculum was functioning well (the theoretical potential for cellulose is 415 NmL $CH_4$/g $VS_{in}$) [29].

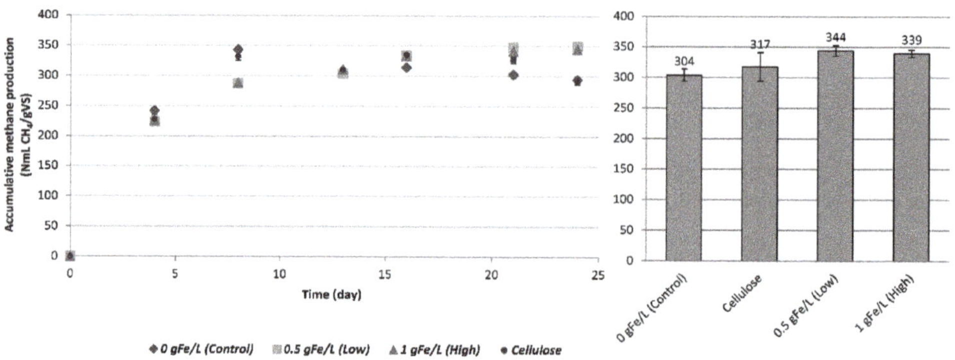


**Figure 2.** The bio-methane potential from the batch experiment for high and low doses of ochre as well as control reactor (no ochre was added). The average values of BMP presented in the figure above (**right**) are based on the last three measurements from the tests shown to the left.

The accumulated methane potentials for the control reactor and reactors with different concentrations of ochre (0.5 and 1 g $Fe^{3+}$/L) are almost similar with a slightly higher potential in the reactors with the ochre addition (around 340 NmL $CH_4$/g $VS_{in}$), which was about 10–15% higher than the methane potential from the control reactor (300 NmL $CH_4$/g $VS_{in}$). The ochre addition did not reduce the methane potential for any of the doses. This is in contrast to the previous studies, where the addition of ochre reduced the methane production by up to 50% [19].

At the end of the batch experiment, the released phosphate was found as shown in Figure 3.

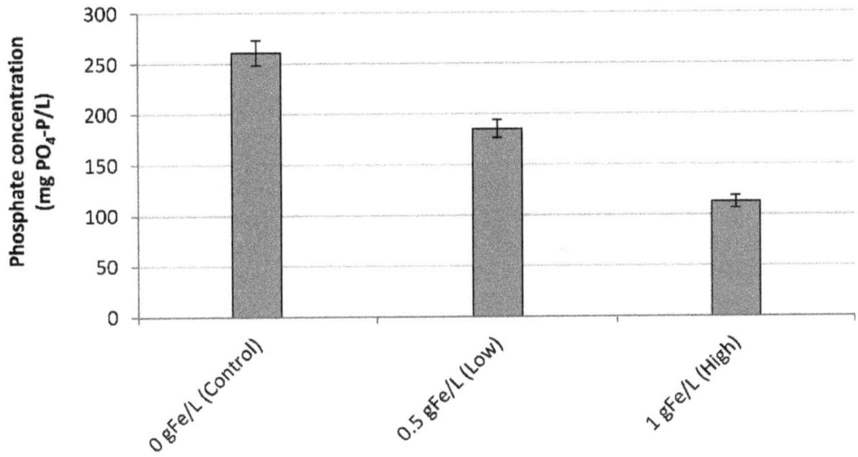


**Figure 3.** The $PO_4$-P concentrations measured in the batch reactors at the end of the batch experiments. The release constituted 261 mg $PO_4$-P/L in the control batch, while reductions of 29% and 57% were observed for the reactors dosed with 0.5 g $Fe^{3+}$/L and 1 g $Fe^{3+}$/L of ochre, respectively.

### 3.3. Continuous Pilot-Scale Experiments

Figure 4 shows the development of hydrogen sulfide during the experimental period.

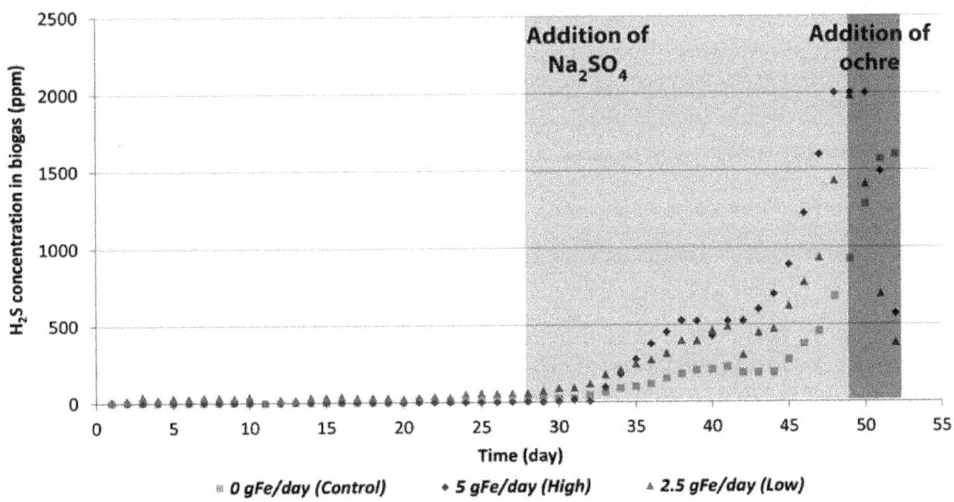


**Figure 4.** The hydrogen sulfide concentration in the produced biogas. Sodium sulfate started to be added in the reactors on day 28 of the experiment in order to provide sulfate as the substrate for the sulfate-reducing bacteria. The dosing of ochre started on day 49 in high (5 g $Fe^{3+}$/day) and low (2.5 g $Fe^{3+}$/day) reactors. No ochre was introduced in the control reactor.

In the start-up period, the $H_2S$ concentration was low (<100 ppm), which was assumed to be the result of having a low sulfate content in the fed sludge. After the addition of sodium sulfate ($Na_2SO_4$) equally to all reactors (to provoke $H_2S$ formation), the concentration increased very rapidly up to >2000 ppm before ochre was added. Both the high (5 g $Fe^{3+}$/d) and low doses (2.5 g $Fe^{3+}$/d) of ochre (corresponding to a molar ratio of 1.6 and 0.8 $Fe^{3+}/S^{2-}$, respectively), resulted in the immediate reduction of hydrogen sulfide ($H_2S$) concentrations (down to 570 ppm), while the control reactor (without ochre addition) still produced biogas with a high $H_2S$-content. It is important to note that sulfide concentrations below 1000 ppm in biogas are recommended in order to guarantee the safe use of the gas heater and a combined heat and power production [6,8,9]. However, volatile fatty acid (VFA) measurements (acetate and propionate) showed that the stability of the reactor with the high ochre dose was affected. The acetate level increased from ~50 mg/L (as COD) before addition, to levels ranging from 230–1700 mg/L after addition. This indicates that ochre in a high dose (5 g $Fe^{3+}$/d or 1.6 mole $Fe^{3+}$/mole $S^{2-}$) was inhibiting the acetate-utilizing microorganism. Previously, it was found that iron inhibits acetic acid conversion during the first 2–4 days of a high iron dose addition [33]. The dosing of ochre was stopped (from day 53), as was the feeding of sludge, to let the reactors recover.

After a short recovery period, the reactors were fed with sludge for 20 days (days 64–84) to eliminate the effect of the ochre dosing ($H_2S$-experiment): After this, a release of phosphates by up to 400 mg $PO_4$-P/L in the sludge liquor was observed. As can be seen in Figure 5, the phosphate concentrations were at the same level in all three reactors, regardless of the previous ochre dosing in two of the reactors. It was also observed that the acetate concentration in the previously high-dosed reactor decreased at the same time. The dosing of ochre (days 85–100) resulted in lower phosphate concentrations in the sludge liquor compared to the control (without ochre dosing).

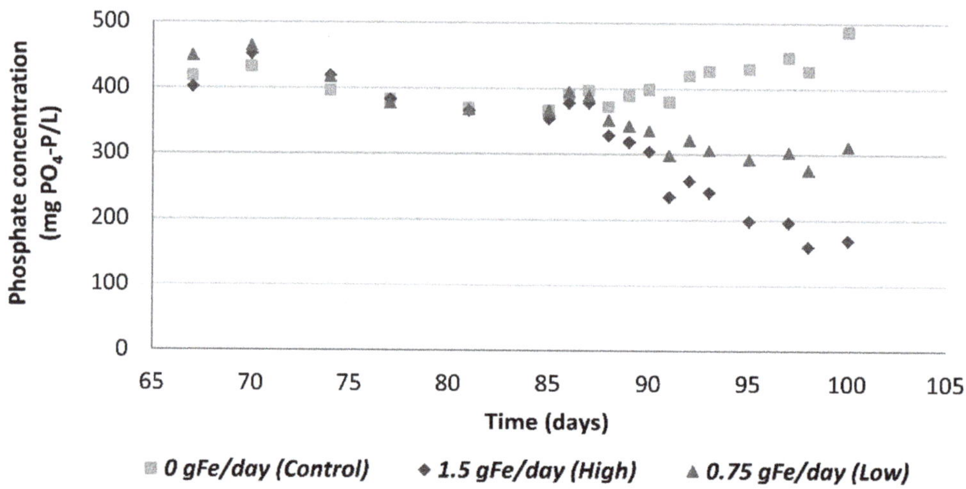


**Figure 5.** The dissolved phosphate concentrations (as mg $PO_4$-P /L) in the continuously operated reactors. The dosing of ochre started on day 85 of the experiment and continued until day 100. The dosage was adjusted to 1.5 g $Fe^{3+}$/day and 0.75 g $Fe^{3+}$/day in the high and low reactors, respectively. No addition of ochre was introduced in the control reactor.

Within a few days, the phosphate concentration levelled out to around 150 mg $PO_4$-P/L in the reactor dosed with 1.5 g $Fe^{3+}$/d (1.5 mole $Fe^{3+}$/mole $P_{released}$). This can be compared to a level of around 300 mg $PO_4$-P/L mg/L for 0.75 g $Fe^{3+}$/d (0.75 mole $Fe^{3+}$/mole $P_{released}$) and of around 450 mg $PO_4$-P/L for the control.

The dosing of minor amounts of ochre can therefore be seen as a cheap method for significantly reducing the internal recycling of phosphate at treatment plants with enhanced biological phosphorus removal. Higher doses of ochre in continuous operation will increase the final sludge production by approximately 11–15%. The iron content in the sludge is expected to be high, around 40–50 g Fe/kg TS, which is similar to the iron content in sludge from wastewater treatments plants with chemical phosphorus removal [12,13,15,16]. Thus, the ochre dose should be optimized in order to keep the extra sludge production to a minimum, but even if the high dose is used the sludge production will in total be similar to that of plants with chemical precipitation.

## 4. Conclusions

We demonstrated the potential for using waste ochre instead of commercial iron to reduce hydrogen sulfide in biogas production and to reduce phosphate release into sludge liquor at WWTPs with enhanced biological phosphorus removal. This use will prevent serious problems in anaerobic digestion and will at the same time solve a significant waste problem resulting from the deposition of ochre from water treatment. However, in order to prevent the inhibition of the methane production process, small doses of ochre should be applied continuously, and the dosing should be optimized in order to reduce extra sludge production. In batch tests, the dosing of ochre (0.5 g $Fe^{3+}$/L and 1 g $Fe^{3+}$/L) showed no inhibition of the biogas production. With the addition of ochre, the accumulated methane potentials in the reactors were 10–15% higher than the methane potential from the control reactor. During the batch test, the phosphate release was reduced by 29% and 57% for the low and high doses, respectively.

During the pilot scale experiment, an immediate drop from 2000 ppm down to 570 ppm of the $H_2S$ concentration in the biogas was seen after dosing with ochre at two different doses: 2.5 g $Fe^{3+}$/d and 5 g $Fe^{3+}$/d. However, the anaerobic conversion process in the reactor with the highest dose (5 g $Fe^{3+}$/d) was inhibited by the ochre, resulting in high acetate concentrations (230–1700 mg/L).

In a second pilot scale experiment, ochre was dosed continuously in smaller amounts (1.5 and 0.75 g $Fe^{3+}$/d) to avoid any inhibition while evaluating the phosphate precipitation. A reduction of phosphates in the sludge liquor (33% and 66% for the low and high doses, respectively) was obtained. The continuous dosing at the highest level that was used in this second test will increase the sludge production by 11–15%.

**Author Contributions:** S.Ö. came up with the basic idea of the project and developed it into experimental plans together with J.l.C.J. All authors participated in the detailed planning of the batch and pilot experiments. S.Ö. did the batch experiment with support from Å.D. and J.l.C.J., S.H. and H.K. did the main part of the pilot experiments with support from the other authors. S.Ö. evaluated the results together with all authors. She wrote the paper with significant input from all authors. All authors have read and agreed to the published version of the manuscript.

**Funding:** The Swedish Institute [Visby grant 00645/2010] and the Swedish Water & Wastewater Association funded the project.

**Acknowledgments:** NSVA (Öresundsverket WWTP) are acknowledged for providing sludge for the experiments and VA SYD (Sjölunda WWTP) for their help with practical problems at the pilot plant.

**Conflicts of Interest:** The authors declare no conflict of interest.

## References

1. Wild, D.; Kisliakova, A.; Siegrist, H. Prediction of recycle phosphorus loads from anaerobic digestion. *Water Res.* **1997**, *31*, 2300–2308. [CrossRef]
2. Pitman, A.R. Management of biological nutrient removal plant sludges—Change the paradigms? *Water Resour.* **1999**, *33*, 1141–1146. [CrossRef]
3. Battistoni, D.; Pavan, M.P.; Prisciandaro, M.; Cecchi, F. Struvite crystallization: A feasible and reliable way to fix phosphorus in anaerobic supernatants. *Water Res.* **2000**, *34*, 3033–3041. [CrossRef]
4. Marti, N.; Bouzas, A.; Seco, A.; Ferrer, J. Struvite precipitation assessment in anaerobic digestion processes. *Chem. Eng. J.* **2008**, *141*, 67–74. [CrossRef]
5. Barat, R.; Bouzas, A.; Martí, N.; Ferrer, J.; Seco, A. Precipitation assessment in wastewater treatment plants operated for biological nutrient removal: A case study in Murcia. *Spain J. Environ. Manag.* **2009**, *90*, 850–857. [CrossRef]
6. Pol, L.W.H.; Lens, P.N.L.; Stams, A.J.M.; Lettinga, G. Anaerobic treatment of sulphate-rich wastewaters. *Biodegradation* **1998**, *9*, 213–224.
7. Visser, A.; Beeksma, I.; van der Zee, F.; Stams, A.J.M.; Lettinga, G. Anaerobic degradation of volatile fatty acids at different sulphate concentrations. *AMicrobiol. Biotechnol.* **1993**, *40*, 549–556. [CrossRef]

8. Appels, L.; Baeyens, J.; Degrève, J.; Dewil, R. Principles and potential of the anaerobic digestion of waste-activated sludge. *Prog. Energy Combust. Sci.* **2008**, *34*, 755–781. [CrossRef]
9. Dhar, B.R.; Youssef, E.; Nakhla, G.; Ray, M.B. Pretreatment of municipal waste activated sludge for volatile sulfur compounds control in anaerobic digestion. *Bioresour. Technol.* **2011**, *102*, 3776–3782. [CrossRef]
10. Mamais, D.; Pitt, P.A.; Cheng, Y.W. Determination of Ferric Chloride Dose to Control Struvite Precipitation in Anaerobic Sludge Digesters. *Water Environ. Res.* **1994**, *66*, 912–918. [CrossRef]
11. Abatzoglou, N.; Boivin, S. A review of biogas purification processes. *Biofuels Bioprod. Biorefening* **2009**, *3*, 42–71. [CrossRef]
12. Smith, J.A.; Carliell-Marquet, C.M. The digestibility of iron-dosed activated sludge. *Bioresour. Technol.* **2008**, *99*, 8585–8592. [CrossRef]
13. Smith, J.A.; Carliell-Marquet, C.M. A novel laboratory method to determine the biogas potential of iron-dosed activated sludge. *Bioresour. Technol.* **2009**, *100*, 1767–1774. [CrossRef]
14. Van Bodegom, P.M.; Scholten, J.C.M.; Stams, A.J.M. Direct inhibition of methanogenesis by ferric iron. *FEMS Microbiol. Ecol.* **2004**, *49*, 261–268. [CrossRef] [PubMed]
15. Novak, J.T.; Verma, N.; Muller, C.D. The role of iron and aluminium in digestion and odor formation. *Water Sci. Technol.* **2007**, *56*, 59–65. [CrossRef]
16. Novak, J.T.; Park, C.M. The effect of iron and aluminium for phosphorus removal on anaerobic digestion and organic sulfur generation. *Water Sci. Technol.* **2010**, *62*, 419–426. [CrossRef]
17. Valentukevičienė, M. Applying backwash water in order to enhance removal of iron and ammonia from spent filters with fresh filter media. *Environ. Prot. Eng.* **2009**, *35*, 135–144.
18. Carlson, I.; Schwertmann, U. Iron and manganese oxides in Finnish ground water treatment plants. *Water Res.* **1987**, *21*, 165–170. [CrossRef]
19. Sharma, A.K.; Thornberg, D.; Andersen, H.R. Application of waterworks sludge in wastewater treatment plants. *Int. J. Environ. Sci. Technol.* **2013**. [CrossRef]
20. Valentukevičienė, M.; Karczmarczyk, A.; Jurkienė, A.; Grigaitytė, A. Research on correlation between nitrogen compounds, iron and manganese concentrations in drinking water supply systems. *J. Environ. Eng. Landsc. Manag.* **2012**. [CrossRef]
21. Rosborg, I. Elevated concentration of iron in drinking water—A potential health risk (Hög koncentration järn i dricksvatten—En potentiell hälsorisk). *J. Water Manag. Res. (VATTEN)* **2009**, *65*, 125–132.
22. Basibuyuk, M.; Kalat, D.G. The use of waterworks sludge for the treatment of vegetable oil refinery industry wastewater. *Environ. Technol.* **2004**, *25*, 373–380. [CrossRef] [PubMed]
23. Kyncl, M. Opportunities for water treatment sludge re-use. *GeoSci. Eng.* **2008**, *1*, 11–22.
24. Paul, E.; Laval, M.L.; Sperandio, M. Excess sludge production and costs due to phosphorus removal. *Environ. Technol.* **2001**, *22*, 1363–1371. [CrossRef] [PubMed]
25. Moelants, N.; Smets, I.Y.; Van Impe, J.F. The potential of an iron rich substrate for phosphorus removal in decentralized wastewater treatment systems. *Sep. Purif. Technol.* **2011**, *77*, 40–45. [CrossRef]
26. Diliūnas, J.; Jurevičius, A.; Zuzevičius, A. Formation of iron compounds in the Quaternary groundwater of Lithuania. *Geology* **2006**, *55*, 66–73.
27. Dauknys, R.; Vabolienė, G.; Valentukevičienė, M.; Rimeika, M. Influence of substrate on biological removal of phosphorus. *Ekologija* **2009**, *55*, 220–225.
28. Jönsson, K.; Jönsson, L.-E.; la Cour Jansen, J. Phosphorus removal without chemicals—Experiences of bio-P in water- and sludge treatment. In Proceedings of the 10th Nordic /NORDIWA Wastewater Conference, Hamar, Norway, 12–14 November 2007.
29. Hansen, T.L.; Schmidt, J.E.; Angelidaki, I.; Marca, E.; la Cour Jansen, J.; Mosbæk, H.; Christensen, T.H. Method for determination of methane potentials of solid organic waste. *Waste Manag.* **2004**, *24*, 393–400. [CrossRef]
30. Davidsson, Å.; Lövstedt, C.; la Cour Jansen, J.; Gruvberger, C.; Aspergen, H. Co-digestion of grease trap sludge and sewage sludge. *Waste Manag.* **2008**, *28*, 986–992. [CrossRef]
31. la Cour Jansen, J.; Gruvberger, C.; Hanner, N.; Aspegren, H.; Svärd, Å. Digestion of sludge and organic waste in the sustainability concept for Malmö, Sweden. *Water Sci. Technol.* **2004**, *49*, 163–169. [CrossRef]

32. Haghighatafshar, S.; Kjerstadius, H.; la Cour Jansen, J.; Davidsson, Å. Management of hydrogen sulfide in anaerobic digestion of enzyme pretreated marine macro-algae. *J. Water Manag. Res. (VATTEN)* **2012**, *68*, 265–274.
33. Hoban, J.; van den Berg, L. Effect of iron on conversion of acetic acid to methane during methanogenic fermentations. *J. ABacteriol.* **1979**, *47*, 153–159. [CrossRef] [PubMed]

*clean technologies*

MDPI

*Article*

# Lomefloxacin—Occurrence in the German River Erft, Its Photo-Induced Elimination, and Assessment of Ecotoxicity

**Melanie Voigt, Benjamin Hentschel, Nina Theiss, Christina Savelsberg, Indra Bartels, Anna Nickisch-Hartfiel and Martin Jaeger ***

Department of Chemistry and ILOC, Niederrhein University of Applied Sciences, D-47798 Krefeld, Germany; Melanie.voigt@hs-niederrhein.de (M.V.); benjovi@ish.de (B.H.); nina.theiss@stud.hn.de (N.T.); Christina.savelsberg@web.de (C.S.); indra.bartels@stud.hn.de (I.B.); anna.nickisch@hs-niederrhein.de (A.N.-H.)

* Correspondence: martin.jaeger@hs-niederrhein.de

Received: 19 December 2019; Accepted: 5 February 2020; Published: 11 February 2020

**Abstract:** Pharmaceuticals in waters represent a worldwide problem of today. Advanced oxidation processes (AOPs) are being researched for elimination of the ecological hazard. Among the substances, the fluoroquinolone antibiotic lomefloxacin was selected for investigation in this study. Lomefloxacin (LOM) was found in the German river Erft. Near and far ultraviolet (UVA, UVC) radiation were used as AOPs and compared for efficiency depending on pH, water matrix, and catalysts. Chemical kinetics description revealed that UVC at pH 8–9 led to the fastest degradation of LOM. The catalysts hydrogen peroxide and titanium dioxide had only limited influence on the degradation rate. Seven novel transformation products were structurally identified by high-resolution higher-order mass spectrometry. Ecotoxicity of the novel and known compounds was assessed by quantitative structure-activity relationship (QSAR) analysis. In addition, irradiation time dependent minimal, and half-maximal inhibitory concentrations (MIC, $IC_{50}$) of LOM solutions were determined and suggested as ecotoxicological hazard indicators. From MIC and kinetic rate constants, the irradiation time required for compound and activity removal could be predicted.

**Keywords:** AOPs; assessment of ecotoxicity; fluoroquinolones; high resolution mass spectrometry; $IC_{50}$; MIC; QSAR

## 1. Introduction

Numerous recent studies have confirmed the presence of various anthropogenic micropollutants in water bodies [1–3]. These micropollutants include pharmaceuticals, personal care products, steroid hormones, and pesticides. The main entry route focuses on wastewater treatment plants (WWTPs), which often fail to completely eliminate these substances through the conventional purification processes. Hence, they enter the aqueous environment where they become ecotoxicologically hazardous. The regular observation of micropollutants during environmental monitoring is often associated with an increasing bacterial resistance formation [1–4]. Observed concentrations range from a few ng $L^{-1}$ to several µg $L^{-1}$. Among the pharmaceuticals, fluoroquinolones are the third largest group of antibiotics in worldwide sales administered to humans and animals and are excreted without metabolization [4,5]. Nakata et al. (2005) reported fluoroquinolone concentrations up to 49 ng $L^{-1}$ both in wastewater treatment effluents and in surface waters and 19 ng $L^{-1}$ in river water [6].

Advanced purification stages for the elimination of these anthropogenic micropollutants are currently being tested worldwide, including chemical catalysts and UV irradiation. Some of these methods, such as UV irradiation, produce OH radicals, which act as strong oxidants with an oxidation potential of 2.8 V [7]. Several fluoroquinolones follow a degradation process according to pseudo-first order kinetics due to OH radical formation [8]. As a disadvantage, these types of advanced oxidation processes (AOPs) may produce degradation or transformation products (TPs) of potentially higher toxicity than the educt. Thus, recent research on the elimination of pharmaceuticals by AOPs extends to toxicity assessment. As suitable measures, quantitative structure-activity relation (QSAR) analysis as well as minimum inhibitory concentration (MIC) and half maximum inhibitory concentration ($IC_{50}$) have been discussed [9–11]. Both methods were compared in this study.

The fluoroquinolone lomefloxacin served as a model compound since this substance was found in various waters in previous studies and also in the nearby River Erft in the course of this study [12,13]. Its irradiation-induced transformation has been described earlier [14,15]. Yet, systematic comparative investigation of near and far ultraviolet (UVA and UVC) irradiation in the presence and absence of hydrogen peroxide and titanium dioxide has not been undertaken. Both reagents were chosen due to their accelerating effect on degradation [16,17]. Time-dependent degradation and/or transformation of educt and known and novel transformation products were monitored using high-performance liquid chromatography and high-resolution mass spectrometry. In parallel, total organic carbon (TOC) was monitored for chemical effectiveness. To consider various aquatic conditions, model water at different pH values, river water, and effluent were used as matrices. As transformation products have often been assumed more ecotoxic than their initial drug substance, irradiation-time dependent MIC and/ or $IC_{50}$ and QSAR analyses were performed to assess the ecotoxicological hazard [18–20]. Mostly, these tools were used to determine the toxicity of new pharmaceuticals, among them fluoroquinolones [21,22]. Only one study could be found concerning photodegradation of fluoroquinolones in combination with QSAR [23]. Based on MIC and chemical kinetics, irradiation times depending on aquatic conditions could be predicted.

## 2. Experiments

### 2.1. Chemicals and Reagents

Lomefloxacin (LOM) was acquired from Molekula (Munich, Germany) and used for all photodegradation experiments. As high performance liquid chromatography (HPLC) eluents, methanol (Merck KGaA, Darmstadt, Germany, LiChrosolv for liquid chromatography) and MilliQ-Water (Millipore System Simplicity 185) were used.

Surface water samples were taken from the German river Erft (Neuss) in summer 2018 downstream of a WWTP and from the river Rhein near Krefeld Uerdingen, see Figure 1. Both samples were checked for the presence of LOM using solid-phase extraction and analyzed with HPLC-ESI-Q-TOF-MS.

**Figure 1.** Sampling of the river Erft (left) and Rhine (right) near a WWTP, (pictures: © 2018 Google, GeoBasis-DE/BKG (© 2009), Google).

### 2.2. Solid-Phase Extraction of River Water Samples

Using solid-phase extraction (SPE), 1 L of a filtered water sample was concentrated to 1 mL. Oasis HLB (Waters, Milford, CT, USA) SPE cartridges were first conditioned with 3 mL of methanol (Merck, Darmstadt, Germany), quenched with 3 mL of MilliQ water, loaded with 1 L of river water, washed with 3 mL of MillliQ water, and then eluted with 3 mL methanol. The substances dissolved in methanol were concentrated to dryness using the rotary evaporator and then taken up again with 1 mL MilliQ water. The sample was cleaned with a syringe filter and then transferred to the HPLC-MS system.

### 2.3. Photodegradation Experiments

Photoinduced degradation experiments were carried out in a 1 L batch reactor (Peschl Ultraviolet GmbH, Mainz, Germany). For irradiation, a UVA light emitting medium pressure mercury lamp (Heraeus, TQ 150, 150 W, Hanau, Germany) was operated with cooling to prevent power associated temperature increase. For UVC irradiation a low-pressure mercury lamp (Heraeus TNN 15/32, 15 W, Hanau, Germany) was operated without need for cooling. The reaction temperature throughout the reactor was 22 ± 2 °C, checked during degradation by means of a thermometer. The lamps were arranged centrically in the reactor. The reactor was filled with 750 and 800 mL aqueous solution with the UVA and UVC lamp inserted. The concentration of LOM was 20 mg $L^{-1}$. The pH was adjusted by adding hydrochloric acid (HCl, 30% Suprapur, Merck KGaA, Darmstadt, Germany) or ammoniacal solution ($NH_3$, approximately 25% Riedel-de Haen, Seelze, Germany). A magnetic stirrer (500 rpm) was used. Irradiation times were 30 min and 10 min for UVA and UVC irradiation, respectively. In previous studies, this exposure time has proven to ensure sufficient degradation [11,24–26].

The warm-up time of the UVA lamp was 2 min vs. 0 min for the UVC lamp. Both lamps emitted polychromatic light with maximum intensities at 313, 365, 405, 437, 547, 578, and 580 nm. The UVC lamp additionally emitted at 185 and 254 nm. The total photon flux in the range between 200 and 500 nm was determined to 3.50 mmol·$min^{-1}$·$L^{-1}$ for the UVA lamp and 2.03 mmol·$min^{-1}$·$L^{-1}$ for the UVC lamp by means of ferrioxalate actinometry according to IUPAC [27,28].

The occurrence of OH radicals on irradiation of water was proven by Electron Spin Resonance (ESR) spectroscopy using spin traps (data not shown) following Sun et al. (1996) and Kochany et al. (1991, 1992) [29–31]. The photocatalytic degradation of LOM was also carried out in the presence of hydrogen peroxide (30% stabilized, Carl Roth, Karlsruhe, Germany) and titanium dioxide (P25 Aeroxide®, Arcos Organics, Geel, Belgium). Hydrogen peroxide was added to the solutions leading to final concentrations of 10 mg $L^{-1}$ and 30 mg $L^{-1}$ in the mixture. Concentrations were determined with Merckoquant test strips before and after irradiation. Further, titanium dioxide was used as a photocatalyst. Addition prior to irradiation led to concentrations of 50 and 100 mg $L^{-1}$ titanium dioxide. The resulting milky solutions were irradiated with UV light.

In addition, degradation experiments were carried out with the effluent of a local WWTP (Krefeld) and from the river Erft (Neuss). The color of the wastewater was yellowish in both cases, the pH values of the two samples were determined to 8. Each sample was spiked with 20 mg $L^{-1}$ LOM. This concentration was chosen to obtain results comparable with those of LOM in MilliQ water.

### 2.4. HPLC-ESI-Q-TOF-MS and HPLC-ESI-IT-MS

For the identification of transformation products at high sensitivity and high reliability, mass spectrometry was used as a detection system after chromatographic separation. Here, time-of-flight (TOF) MS was used for highly accurate m/z determination. Multiple fragmentation for structure elucidation and verification was achieved through ion-trap MS. Samples were analyzed using a high-performance liquid chromatography system (HPLC, Agilent 1100) combined with an ion-trap mass spectrometer (IT-MS, Thermo Finnigan LXQ, Waltham, MA, USA) to yield $MS^n$. A reversed-phase column 3-Amides C-18, 150 mm × 2.00 mm, 3 μm (Polaris Agilent) was used for chromatographic separation. Temperature was kept at 40 °C and chromatography was performed isocratically within 15 min using MilliQ water (90%) and methanol (10%) with 0.1% formic acid each (Merck KGaA, EMSURE® ACS, Darmstadt, Germany, 98–100%) as eluents. The flow rate was set to 0.3 mL/min for all measurements. An electrospray ionization (ESI) source was attached to the IT-MS. For all experiments, the positive mode was used. The gas flow was adjusted to 15 L $min^{-1}$, the capillary temperature was set to 300 °C and the capillary voltage amounted to 47 V. The mass range was scanned from 100 to 1000 m/z. Tandem or higher MS experiments were initiated using the software's auto $MS^n$ mode using a threshold trigger. Mass spectrometer and HPLC system were controlled via XCalibur 2.0 running on a personal computer under Windows XP. Chromatograms and mass spectra were processed using the same software. Accurate masses were obtained using an ESI-Q-TOF mass spectrometer (Agilent 6530). A dual jet stream (ESI) source was used in positive mode. The collision gas flow was set to 8 L $min^{-1}$, the gas temperature was 300 °C and the fragmentor voltage 175 V. The instruments were controlled via MassHunter Workstation B.06.00 running on a personal computer under Windows 7 Professional. Data were processed using the same software.

### 2.5. TOC

The total organic carbon (TOC) was determined using a TOC analyzer (TOC-V CSN Series Shimadzu, Kyoto, Japan). Calibration was performed between 2 and 20 mg $L^{-1}$ carbon using solutions of sucrose. The pH of all samples was adjusted to 1–3 using concentrated sulfuric acid. Subsequently, potentially resulting gases in the samples were removed by ultrasonification for 15 min. Samples were taken before irradiation and during irradiation after 15, 30, 45, and 60 min and every hour for a total of 7 h. Injection volume was 150 µL for each sample. The device was controlled and results evaluated using TOC-Control software (Shimadzu, Kyoto, Japan). Each sample was measured in triplicates. Mean values yielded the concentrations. Prior to measurements, the TOC analyzer was equilibrated with the new sample to remove residual carbon. In addition, between each LOM sample acidified and degassed MilliQ water was measured.

### 2.6. Kinetic Analysis and Determination of Quantum Efficiency of the Photodegradation

Kinetic analysis of degradation profiles followed first-order or pseudo first-order kinetic models, consecutive and subsequent follow-up reaction models, cf. Equation (1), according to previous studies and theory [26,32–37]. The complete kinetic description is represented in the Supplementary Information.

$$A \xrightarrow{k_1} B + \ldots \tag{1}$$

Substance $A$ reacts with the reaction rate constant $k_1$ yielding an unknown number of products. The product can continue to react. The sequence can be understood as a consecutive or subsequent follow-up reaction. Rigorous mathematical treatment returns the time-dependent concentration $c$ of educt $A$, with $c_A$ the actual and $c_{A_0}$ the initial concentration of $A$, and the time $t$, cf. Equation (2).

$$c_A = c_{A_0} \cdot e^{-k_1 t} \tag{2}$$

From mass area-under-the-curve versus irradiation time graphs, degradation curves were obtained. The software MatLab R2017a (MathWorks, Natick, MA, USA) containing the curve fitting toolbox was used to compute the kinetic profiles of the degradation intermediates and products according to Equations (2)–(4) (see also Supplementary Information).

### 2.7. Determination of Quantum Efficiency

The quantum efficiency may be defined as the degradation rate divided by the number of photons absorbed [29]. The quantum efficiency $\Phi_{254}$ at 254 nm was calculated using Equation (3) [38].

$$\Phi_{254} = \frac{k}{2.303 l \varepsilon_{254} I_{0,254}} \tag{3}$$

where $k$ ($min^{-1}$) is the reaction rate constant of the degradation A→B, $l$ is the reactor width of 3 cm, $\varepsilon_{254}$ is the molar extinction coefficient in L $mol^{-1}$ $cm^{-1}$ and $I_{0,254}$ the photon fluence rate in mmol $min^{-1}$ $L^{-1}$.

*2.8. Assessment of Ecotoxicology*

The assay method for the determination of MIC values of LOM followed the protocol by Wiegand, Hilpert, and Hancock [39]. The assay protocol is also given in ISO 20776-1:2007. As ubiquitous bacteria, *Pseudomonas fluorescens* (DSM-No. 50090) and *Bacillus subtilis* (DSM-No. 10) were selected. Bacterial growth was monitored before and after UV irradiation [40]. For more details see Supplemental Information. The calculation of $IC_{50}$ was performed on the basis of the plot of bacterial growth versus compound concentration.

The time $t_{act}$ according to Equation (4) represents the decline in activity an antibiotic possesses against bacteria. It can be calculated from kinetic parameters obtained from degradation experiments, i.e., the and the rate constants $k$ and the initial compound concentration $c_{A0}$, and the MIC values (see above) [36].

$$t_{act} = -\frac{\ln\left(\frac{MIC}{c_{A_0}}\right)}{k_1} \tag{4}$$

The software distributions T.E.S.T. and OECD QSAR Toolbox were applied for QSAR analyses [41,42]. Within T.E.S.T, the module QSAR was selected. Compound structures were drawn using ACD/ChemSketch 2016.1.1 (ACDLabs, Inc., Toronto, ON, Canada) and imported into the software T.E.S.T and QSAR toolbox. As relevant target organisms, *Daphnia magna* $LC_{50}$ (48 h) in mg $L^{-1}$, fathead minnow $LC_{50}$ (96 h), 'Photoinduced Toxicity on *D. magna*', and 'Mortality $LC_{50}$ (48 h) of branchiopoda' for non-specified test organisms were chosen due to their presence in the aquatic environment.

## 3. Results and Discussion

Selected physico-, photo-chemical, and mass spectrometric properties of LOM are collected in Table 1.

**Table 1.** Properties of Lomefloxacin (LOM).

| Fluoroquinolone | Lomefloxacin |
|---|---|
| Chemical Structure | 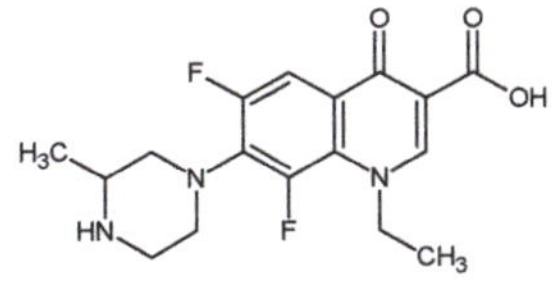 |
| Molecular Formula | $C_{17}H_{19}F_2N_3O_3$ |
| CAS Reg. No. | 98079-51-7 |
| pKa (1) | 6.3; 9.0 |
| $[M+H]^+$ (exact) | 352.147 |
| $[M+H]^+$ (accurate) (2) | 352.147 |
| MS/MS (3) | 308 |
| MS/MS/MS (3) | 265; 288 |
| λ ($\varepsilon_{max}$)/nm at pH 3 | 287 |
| $\varepsilon_{max}$/L·$mol^{-1}$ at pH 3 | 37624 |
| λ ($\varepsilon_{max}$)/nm at pH 7 | 282 |
| $\varepsilon_{max}$/L·$mol^{-1}$ at pH 7 | 32548 |
| λ ($\varepsilon_{max}$)/nm at pH 9 | 282 |
| $\varepsilon_{max}$/L·$mol^{-1}$ at pH 9 | 32181 |

(1) Babic et al.; (2) Determined using Q-TOF MS; (3) Determined using IT MS.

In neutral solutions, fluoroquinolones occur as zwitterions, as can be recognized from their pKa values. This property exercises influence on the photoinduced degradation through directing the chemical transformation and through the UV absorption efficiency, which depends on the pH as well, see Figure 2a. The spectral range of the irradiation also affects the quantum efficiencies. Hence, UVC or UVA lamps used in this study may lead to different transformation pathways and products. In order to predict conditions for efficient degradation of LOM, absorption spectra were recorded at different pH values and compared with the emission spectra of the lamps, see Figure 2b.

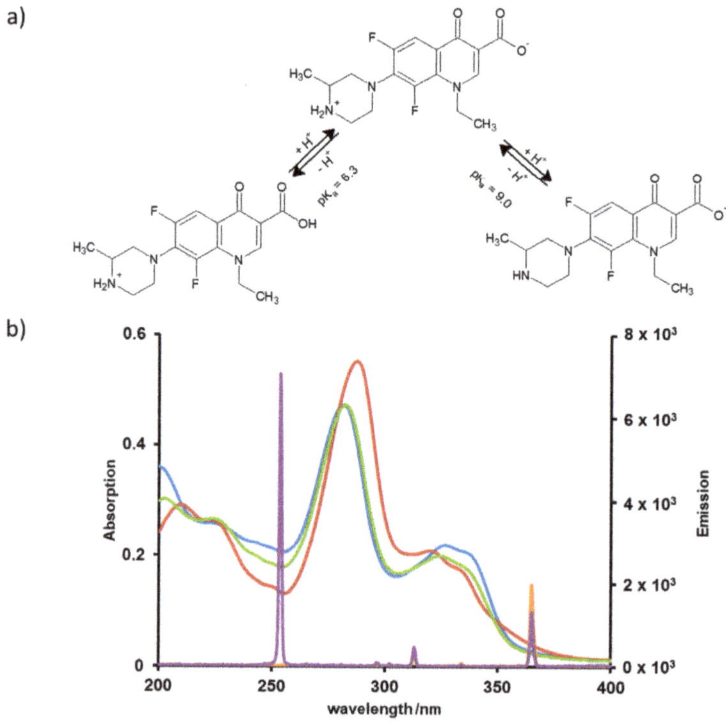


**Figure 2.** (**a**) pH dependent protonation and deprotonation of LOM; (**b**) Absorption spectra of LOM at pH 3 (red), pH 7 (green) and pH 9 (blue) compared to the emission spectra of the UVA (orange) and UVC lamp (purple).

### 3.1. *Occurrence of Lomefloxacin in Surface Water*

The presence of LOM is a worldwide problem, and thus, also in Germany. A sample from the River Erft showed the occurrence of LOM, whereby no LOM was observed in the sample from the River Rhine in summer 2018. The findings were confirmed through comparison of chromatographic retention times and the MS/MS fragmentation pattern with a LOM reference sample, see Figure 3. The accurate mass observed agreed also perfectly with the theoretical monoisotopic mass, cf. Table 1. Further substances were found in the river, such as the $b_1$ receptor blocker metoprolol. The remaining substances found in both river water samples were not considered in this study.

Samples containing LOM were taken from the River Erft immediately after a sewage treatment plant, while in samples collected upstream of the plant showed no presence of LOM. In agreement with previous studies, WWTPs seem to concentrate LOM and act as entry path into the aquatic environment [43,44]. Findings as this underline the need for a fourth purification stage based on AOPs or at least advanced filtering systems. In this respect, the photo-induced degradation of LOM is investigated.

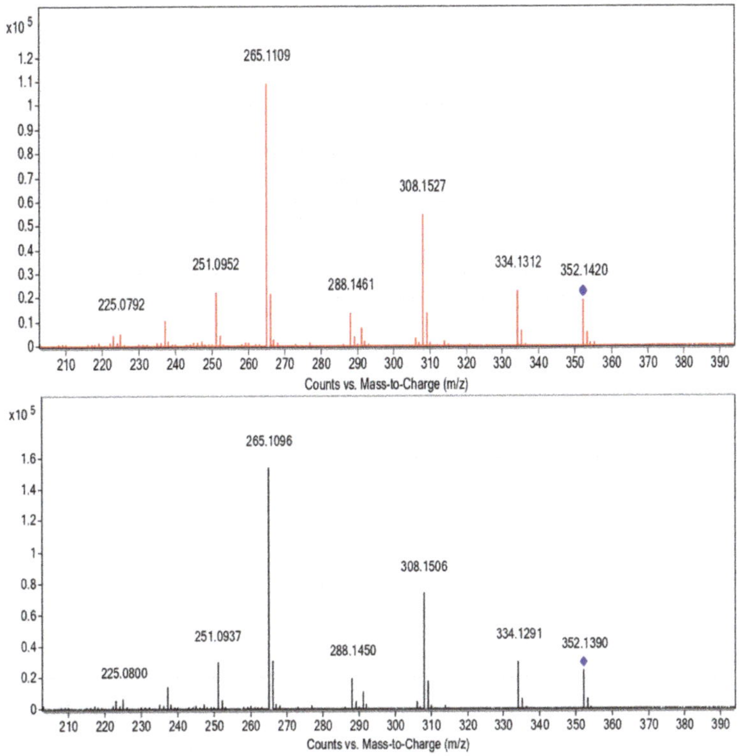


**Figure 3.** MS/MS spectra of the quasi molecular ion from LOM dissolved in MilliQ-water (bottom) and from samples of the River Erft (top).

## 3.2. *Photoinduced Degradation of Lomefloxacin—Kinetic of Degradation*

The influence of UVA and UVC radiation, pH values, and model and surface water matrices on the photoinduced degradation were analyzed. Examples are given in Figure 4. Photoinduced transformation and degradation were monitored using HPLC-MS techniques and plotted as concentration or area-under-the-curve mass signals versus time (*c*-*t*).

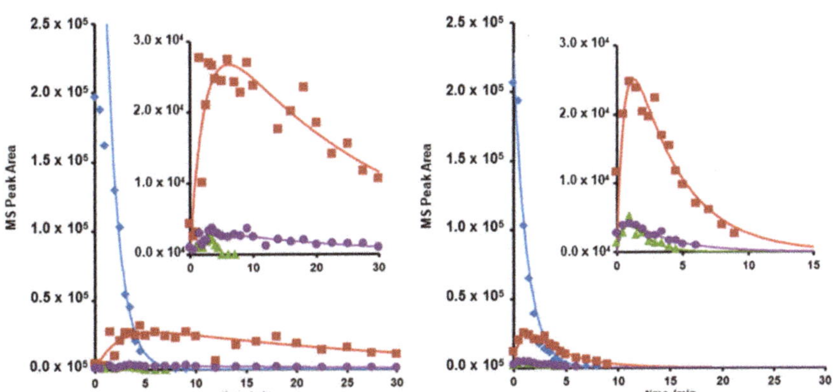


**Figure 4.** *c*-*t* curves of LOM (♦) and photodegradates with m/z = 308 (▲), 332 (■) and 350 (•) under UVA (left) and UVC irradiation (right) at pH = 4–6.

The *c-t* curves obtained could be best described by first or pseudo-first order kinetics in agreement with previous studies [25,26,36]. The results are presented in Table 2 as kinetic rate constants and quantum efficiencies.

**Table 2.** Comparison of the kinetic rate constants, half-lives, and quantum efficiencies at the wavelengths 254 nm, 313 nm and 365 nm obtained for UVA and UVC irradiated solutions of 20 mg $L^{-1}$ LOM in MilliQ, river Erft water, and effluent.

| Radiation Source | Water Matrix | pH | $k_1$/min$^{-1}$ | $t_{1/2}$/min | $\Phi_{254\ nm}$ | $\Phi_{313\ nm}$ | $\Phi_{365\ nm}$ |
|---|---|---|---|---|---|---|---|
| UVA | MilliQ | 3–4 | 0.96 | 0.72 | - | 0.63 | 0.29 |
| | MilliQ | 4–6 | 0.82 | 0.85 | - | 0.52 | 0.32 |
| | MilliQ | 6–7 | 1.22 | 0.57 | - | 0.73 | 0.45 |
| | MilliQ | 8–9 | 1.99 | 0.35 | - | 1.30 | 0.79 |
| UVC | MilliQ | 3–4 | 0.57 | 1.21 | 0.02 | 0.13 | 0.39 |
| | MilliQ | 4–6 | 0.80 | 0.87 | 0.02 | 0.15 | 0.62 |
| | MilliQ | 6–7 | 2.83 | 0.24 | 0.09 | 0.57 | 2.40 |
| | MilliQ | 8–9 | 2.04 | 0.34 | 0.06 | 0.45 | 1.90 |
| | River Erft | 6–7 | 1.49 | 0.47 | | | |
| | Effluent WWTP | 8 | 1.47 | 0.47 | | | |

Radiation range, pH, and matrix were found to influence the degradation rate. While the degradation of LOM in acidic milieu was slowest, it could be quadrupled by increasing the pH. The water matrix also affects the degradation rate. In MilliQ water, the degradation is faster compared to effluent and river water. In environmental water matrices, a large variety of compounds are present and decrease the amount of light reaching LOM which induces its transformation or degradation. The effect of pH on the rate constants may be twofold. Firstly, LOM itself may be degraded faster in its zwitterionic or anionic state, like in neutral and alkaline milieu. Secondly, the photoinduced formation of hydroxyl radicals contributes to the degradation and is pH dependent. The presence of hydroxide hampers hydroxyl radical induced degradation, hence would decrease the part of the rate constant originating from this mechanism. Yet, hydroxide can act as a reagent itself and would lead to an increase of the rate constant. As can be seen from Table 2, quantum efficiencies exceeding 1 suggest more than one mechanism in action. In conclusion, the obviously best conditions for the removal of LOM were UVC irradiation and a pH between neutral and slightly alkaline.

Among AOPs, photocatalysts, such as titanium dioxide in combination with UVA irradiation or hydrogen peroxide in combination with UVC irradiation, were often used to accelerate the degradation of many anthropogenic micropollutants [7,45–47]. Yet, no significant differences were observed between presence and absence of these photocatalysts in this study, see Table 3. The sole exception occurred when the degradation was accelerated in the presence of hydrogen peroxide under UVC irradiation at pH 3–4. The lack of efficiency of hydrogen peroxide was traced back to saturation effects [7,48], while the presence of titanium dioxide was assumed to cause light scattering hence reducing the amount of light for direct degradation [49]. Since effluents most often show pH values in the neutral or weak alkaline range, no need for the application of hydrogen peroxide or titanium dioxide was indicated.

To this point, only the induced degradation of LOM was considered. Since organic decomposition products were formed, it was also interesting to monitor TOC with irradiation time. Figure 5 shows the irradiation time dependent TOC of LOM in river Erft water and MilliQ water at an initial concentration of 20 mg $L^{-1}$ and under UVC irradiation.

The initial TOC in river water was about 6.5 times higher than that of MilliQ water containing 20 mg $L^{-1}$ of LOM, since the river water contained a variety of organic substances. After 7 h of UVC irradiation, the TOC in both samples was reduced by 10 mg $L^{-1}$. As the initial TOC of the MilliQ water sample amounted to 10 mg $L^{-1}$, complete mineralization was achieved. In contrast, the irradiation time of the river Erft water sample would have had to be significantly extended for complete mineralization.

It is obvious that TOC is no suitable measure for the removal of LOM or other pharmaceuticals as surface waters or effluents carry innumerable organic compounds that react differently to UV exposure. Nevertheless, it is reasonable to assume that LOM degradation proceeded via transformation products that were analyzed and identified using HPLC-ESI-Q-TOF-MS and HPLC- ESI-IT-MS as described in the following section.

**Table 3.** Determined rate constants and the resulting half-lives with UVA and UVC irradiation and $H_2O_2$ or $TiO_2$ addition.

| pH | $H_2O_2$ /mg·L$^{-1}$ | $TiO_2$ P25/mg·L$^{-1}$ | UVA $k$/min$^{-1}$ | UVA $t_{1/2}$/min | UVC $k$/min$^{-1}$ | UVC $t_{1/2}$/min |
|---|---|---|---|---|---|---|
| 3–4 | 0 | 0 | 0.96 | 0.72 | 0.57 | 1.21 |
| | 0 | 50 | 1.71 | 0.41 | | |
| | 0 | 100 | 0.79 | 0.88 | | |
| | 10 | 0 | | | 0.93 | 0.75 |
| | 30 | 0 | | | 1.03 | 0.68 |
| 6–7 | 0 | 0 | 1.22 | 0.57 | 2.83 | 0.24 |
| | 0 | 50 | 1.07 | 0.65 | | |
| | 0 | 100 | 1.42 | 0.49 | | |
| | 10 | 0 | | | 2.10 | 0.33 |
| | 30 | 0 | | | 1.84 | 0.38 |
| 8–9 | 0 | 0 | 1.99 | 0.35 | 2.04 | 0.34 |
| | 0 | 50 | 0.80 | 0.87 | | |
| | 0 | 100 | 1.05 | 0.66 | | |
| | 10 | 0 | | | 1.30 | 0.53 |
| | 30 | 0 | | | 2.34 | 0.30 |

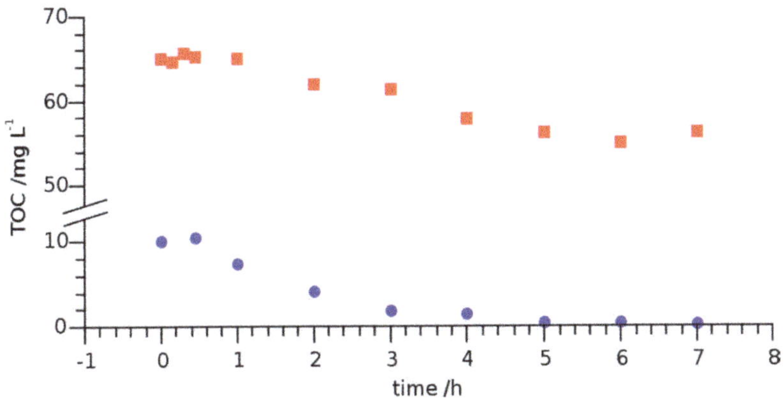


**Figure 5.** Total organic carbon (TOC) of LOM dissolved in MilliQ-water (•) and river Erft water (■) at initial concentration of 20 mg L$^{-1}$ under UVC irradiation.

### 3.3. *Photoinduced Degradation Products of Lomefloxacin*

During the degradation experiments, both known and unknown compounds were identified by HPLC-ESI-IT-MS and/or HPLC-ESI-Q-TOF-MS. An overview of the observed substances is shown in Table 4.

**Table 4.** Photoinduced transformation products and their purposed structure. Identification using MS and MS/MS experiments.

| $[M + H]^+$ (Theoretical) | $[M + H]^+$ (Observed) | $MS^2$ | $MS^3$ | Proposed Structure | Reference |
|---|---|---|---|---|---|
| 352.147<br>Lomefloxacin | 352.147 | 308 | 265 | 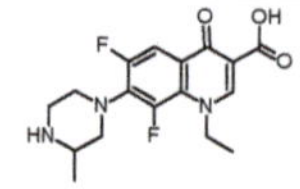 | |
| 294.125 | 294.123 | 276 | | 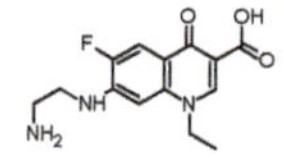 | this study |
| 300.134 | 300.148 | | | 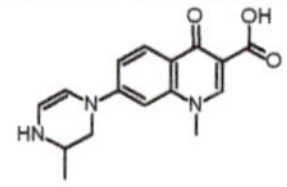 | this study |
| 304.129 | | | | 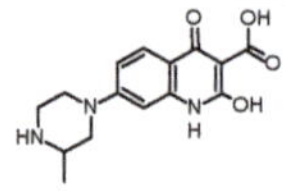 | this study |
| 308.140 | 308.140 | 290 | 270 | 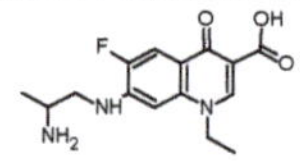 | [50,51] |
| 322.120 | | | | 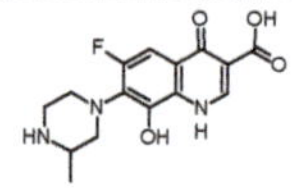 | this study |
| 324.115 | 324.1468 | 281 | 253 | 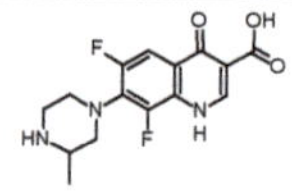 | this study |
| 332.140 | 332.136 | 288 | 245 | 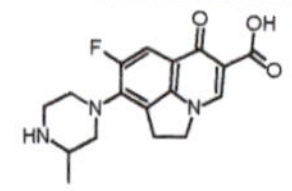 | [50–53] |
| 346.140 | | | | 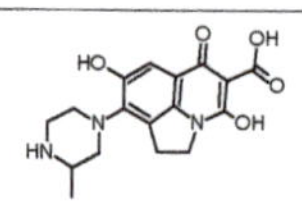 | this study |
| | | | | 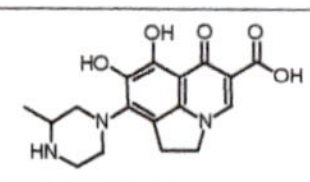 | |
| 348.135 | 348.135 | 330 | 304 | 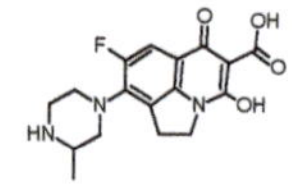 | [51] |
| 348.135 | | | | 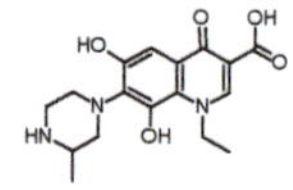 | |
| 350.151 | 350.144 | | | 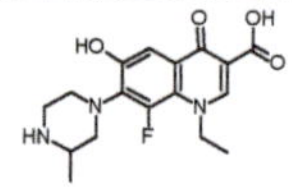 | [51] |

**Table 4.** *Cont.*

| [M + H]$^+$ (Theoretical) | [M + H]$^+$ (Observed) | MS$^2$ | MS$^3$ | Proposed Structure | Reference |
|---|---|---|---|---|---|
| 350.151 | | 306 | 263 | 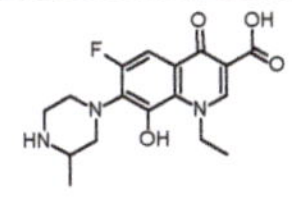 | |
| 350.151 | | 332 | 267 | 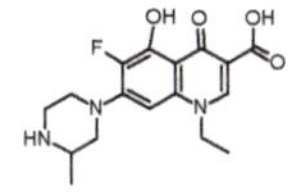 | |
| 364.150 | | 346 | 195 | 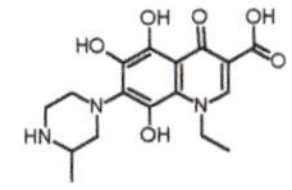 | this study |
| 364.150 | | | | 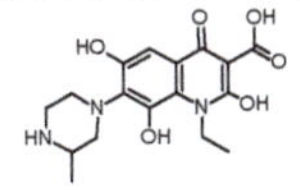 | |
| 368.142 | 368.111 | | | 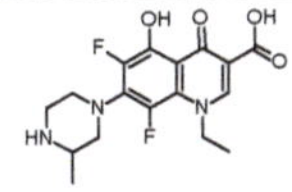 | [54] |

Some of the photodegradates shown in Table 4 were preferably formed on UVA irradiation, others exclusively on UVC irradiation. The latter gave rise to many more transformation products. In particular, the effect of hydroxyl group substitution was recognized when using different radiation sources. Alkaline pH and UVC irradiation yielded relatively many photodegradates with hydroxyl substituents, whereas these products could not be identified on UVA irradiation, see green marks in Figure 6. Photoinduced deconstruction of the piperazine moiety seemed independent of pH, see orange marks in Figure 6. The same observation was made for reactions of the ethyl substituent on the quinolone nitrogen, cf. purple marks in Figure 6. Although it cannot be ruled out that different mechanisms might lead to identical transformation products, it could be concluded that pH independent UVA and UVC photoinduced chemistry led to the fragmentation of the piperazine ring. The ethyl substituent reactions might stem from hydroxyl radical and hydroxide chemistry depending on pH, as hydroxide acts as radical scavenger for hydroxyl radicals. At last, fluorine substitution by and addition of hydroxyl groups might also be due to both mechanisms, depending on pH. An overview of the observed transformation products is given in Figure 6. As indicated in Table 4, seven not previously reported products were proposed based on MS$^2$ and MS$^3$ experiments.

Following kinetic analysis of the *c-t* diagrams, see Figure 4, most of the transformation products could be described as intermediate products. While only the intermediate with m/z = 332 due to UVA irradiation possessed a life-time well above 30 min, all photoproducts due to UVC irradiation were eliminated within 15 min, cf. Figure 4. Under suitable neutral and slightly alkaline conditions, UVC irradiation led to faster elimination of LOM and its photoproducts than UVA irradiation.

### 3.4. Assessment of Ecotoxicity

To assess the ecotoxicity of the resulting products and intermediates of the photoinduced elimination of LOM, $IC_{50}$, and MIC values were determined. The MIC value of LOM against *B. subtilis* was found 0.17 μg·mL$^{-1}$, and 0.92 μg·mL$^{-1}$ against *P. fluorescens*. The $IC_{50}$ value of LOM against *B. subtilis* was 0.09 μg·mL$^{-1}$ and 0.22 μg·mL$^{-1}$ against *P. fluorescens*. As discussed in previous studies, $IC_{50}$ values can be determined more accurately than MIC values due to their graphical representation [9]. Yet, an error of a factor of 2 is negligible due to the dilution assay format. A factor of 4 between $IC_{50}$

and MIC in the case of *P. fluorescens* is rather large. Example curves for $IC_{50}$ determination are given in Figure 7.

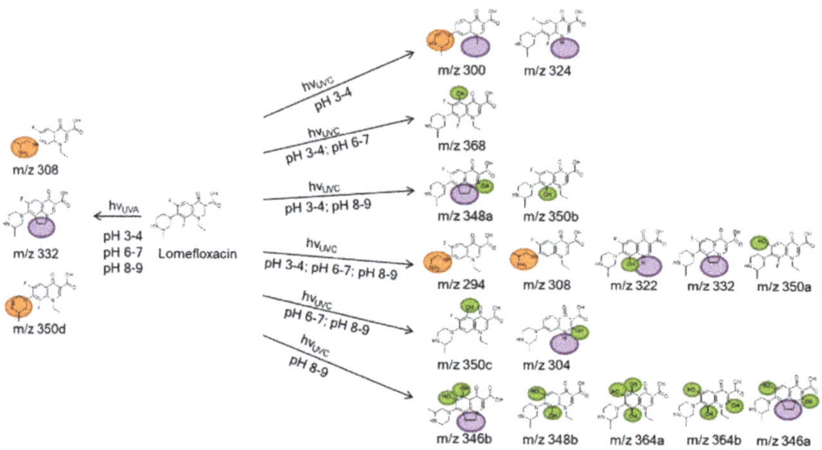

**Figure 6.** Identified photodegradation products of LOM formed by OH radical addition (green), alteration of the piperazine moiety (orange), the nitrogen substituent (purple), and the quinolone ring (blue).

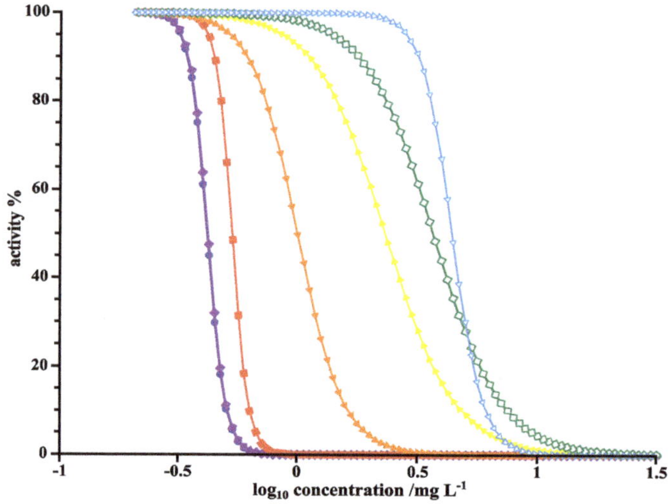

**Figure 7.** Activity-concentration curves of LOM against *P. fluorescens* for the determination of $IC_{50}$ values at UVA irradiation times of 0 min (•), 1 min (♦), 2.5 min (■), 3.5 min (▼), 5 min (▲), 7 min (◊), and 30 min (∇).

As expected, the longer the LOM solution was irradiated with UVA light, the higher the $IC_{50}$ grew and the least activity remained. The comparison between MIC and $IC_{50}$ values depending on time of irradiation by UVA and UVC is shown in Table 5.

**Table 5.** MIC values of LOM against *P. fluorescens* and *B. subtilis* depending on UVA and UVC irradiation times.

| Irradiation Time/min | *P. fluorescens*/µg·mL$^{-1}$ | | | | *B. subtilis*/µg·mL$^{-1}$ | | | |
|---|---|---|---|---|---|---|---|---|
| Irradiation Source | UVA | | UVC | | UVA | | UVC | |
| | MIC | $IC_{50}$ [(a)] | MIC | $IC_{50}$ [(a)] | MIC | $IC_{50}$ [(a)] | MIC | $IC_{50}$ [(a)] |
| 0 | <0.55 | 0.41 | <0.55 | 0.56 | 0.28 | 0.14 | 0.28 | 0.14 |
| 1 | <0.55 | 0.42 | 1.11 | 0.80 | 0.28 | 0.17 | 0.55 | 0.22 |
| 2.5 | 1.10 | 0.53 | 4.44 | 2.34 | 0.55 | 0.13 | 2.22 | 0.51 |
| 3.5 | 2.21 | 1.00 | 8.88 | 0.95 | 1.10 | 0.26 | >2.22 | 0.21 |
| 5 | 4.42 | 2.31 | >8.88 | 2.59 | 2.21 | 0.44 | >2.22 | 0.60 |
| 7 | 8.83 | 3.68 | >8.88 | 1.76 | >2.21 | 0.18 | >2.22 | 0.45 |
| 30 | >8.83 | 4.41 | >8.88 | 1.27 | >2.21 | 0.65 | >2.22 | 0.22 |

[(a)] $IC_{50}$ values were erranous as insufficient data points were obtained due to too low activity of irradiated solutions.

In general, activity decreased with irradiation times where UVC irradiation appeared more effective than UVA, as can be seen from the higher MIC values at equal irradiation times. Since both bacteria represent Gram-positive and Gram-negative species ubiquitous in the aquatic environment, the lack of activity of a compound or compound mixture has been proposed as means to assess the ecotoxicity [11]. The advantage of the assay is its ease of use. While QSAR analyses often rely on the knowledge of the exact structures of the transformation products after AOP application, MIC or $IC_{50}$ values represent sum parameters and therefore reflect the activity in total, hence answering indirectly to an assumption that transformation products might be more ecotoxic than the initial drug itself.

To address the issue whether transformation products are less or more toxic than their parent drug, QSAR analyses represent a common approach, since the synthesis of the suggested products is usually expensive, and no corresponding standards were available in this case. QSAR analysis was carried out on the basis of the chemical structures of the known and newly identified products of LOM, see Table 4. The results for selected species are presented in Table 6.

Most of the photoproducts observed in this work were predicted less toxic by the QSAR methods. The result was expected since the groups known to be responsible for the efficacy of fluoroquinolones were altered or deconstructed through irradiation or photoinduced chemistry. Differences in predictions against the same organisms were due to different programs or databases contained therein. Values for *Daphnia magna* were not obtained from the QSAR toolbox. A direct comparison could be achieved for fathead minnow. Here, the values differed by several orders of magnitude, and the toxicity was predicted differently for several product structures, cf. Table 6. This can be traced back to different databases and calibration models contained in the two programs. The prediction and computation of t.e.s.t. software is based on the database of EUCAST, while the QSAR toolbox calculation is based on the data of the European Chemicals Bureau. A literature search for reference data remained unsuccessful. Despite the different absolute values, both predictions were consistent with reference to the parent drug. Photoproducts were mostly predicted potentially less toxic. An exception was the product with $m/z$ = 324, which strongly resembled the educt after elimination of the ethyl substituent. Yet, QSAR analyses do not take concentrations into account, hence do not reflect the ecotoxicological activity of a mixture. In this respect, MIC and $IC_{50}$ values provide total activity information, although the extent of relevance of the model bacteria to the aquatic environment has not been proven yet.

When irradiation is considered for removal of pharmaceuticals before entering the environment, degradation rate constants in combination with MIC- or $IC_{50}$- values allow to determine the irradiation time according to Equation (4). The equation relates activity against bacteria with compound transformation or degradation. Hence, the time to remove potential ecotoxicological hazard from a sample containing LOM could be estimated. The values obtained for LOM samples in MilliQ water and effluent exposed to UVC irradiation are shown in Table 7.

**Table 6.** QSAR analysis of the phototransformation products of LOM with less (green) and more hazardous (red) photoproducts as compared to the parent compound. The photoproducts are indicated by their *m/z* values. The activities against the organisms are given as concentration.

| Substance | *Daphnia magna* $LC_{50}$ (48 h) mg/L [a] | Fathead Minnow $LC_{50}$ (96 h) mg/L [a] | Fathead Minnow $LC_{50}$ (96 h) mg/L [b,c] | Actinopterygii $LC_{50}$ (96 h) mg/L [b,d] | Branchiopoda $LC_{50}$ (48 h) mg/L [b,d] |
|---|---|---|---|---|---|
| Lomefloxacin | 7.87 | 0.42 | $3.15 \times 10^2$ | $2.51 \times 10^6$ | $8.52 \times 10^2$ |
| 294 | 9.62 | 1.95 | $2.77 \times 10^3$ | $3.69 \times 10^8$ | $9.94 \times 10^3$ |
| 300 | 63.58 | 2.10 | $1.76 \times 10^3$ | $1.12 \times 10^8$ | $3.37 \times 10^3$ |
| 304 | 24.82 | 3.16 | $7.80 \times 10^3$ | $3.83 \times 10^9$ | $1.88 \times 10^4$ |
| 308 | 9.80 | 1.50 | $1.54 \times 10^3$ | $1.15 \times 10^8$ | $3.46 \times 10^3$ |
| 322 | 15.04 | 1.39 | $2.97 \times 10^3$ | $7.20 \times 10^8$ | $8.61 \times 10^3$ |
| 324 | 4.28 | 0.40 | $1.09 \times 10^3$ | $1.02 \times 10^8$ | $3.35 \times 10^3$ |
| 332 | 16.38 | 0.70 | $5.00 \times 10^2$ | $1.07 \times 10^7$ | $1.14 \times 10^3$ |
| 346a | 17.20 | 0.63 | $4.73 \times 10^3$ | $1.03 \times 10^9$ | $1.06 \times 10^4$ |
| 346b | 13.67 | 0.22 | $9.50 \times 10^2$ | $3.44 \times 10^7$ | $2.05 \times 10^3$ |
| 348a | 12.18 | 0.44 | $1.44 \times 10^3$ | $1.46 \times 10^8$ | $4.14 \times 10^3$ |
| 348b | 17.07 | 0.90 | $2.62 \times 10^3$ | $2.76 \times 10^8$ | $5.63 \times 10^3$ |
| 350a | 13.05 | 0.70 | $8.78 \times 10^2$ | $3.91 \times 10^7$ | $2.19 \times 10^3$ |
| 350b | 15.62 | 0.36 | $9.06 \times 10^2$ | $3.91 \times 10^7$ | $2.19 \times 10^3$ |
| 350c | 15.95 | 0.69 | $5.21 \times 10^2$ | $1.13 \times 10^7$ | $1.20 \times 10^3$ |
| 350d | 9.22 | 0.09 | $1.99 \times 10^1$ | $7.95 \times 10^2$ | $1.36 \times 10^2$ |
| 364a | 9.46 | 0.67 | $1.51 \times 10^3$ | $1.22 \times 10^8$ | $3.87 \times 10^3$ |
| 364b | 15.67 | 0.78 | $8.52 \times 10^3$ | $3.64 \times 10^9$ | $2.01 \times 10^4$ |
| 368 | 10.75 | 0.42 | $3.73 \times 10^2$ | $6.69 \times 10^6$ | $9.55 \times 10^2$ |

[a] T.E.S.T; [b] QSAR Toolbox; [c] *Pimephales promelas*; [d] unknown testorganisms.

**Table 7.** Times $t_{akt}$ after which the effectiveness of LOM against *P. fluorescens* and *B. subtilis* subsides on UVC irradiation, depending on pH and matrix.

| pH | water | $t_{akt}$ (*P. fluorescens*)/min | $t_{akt}$ (*B. subtilis*)/min |
|---|---|---|---|
| 3–4 | MilliQ | 5.40 | 8.36 |
| 6–7 | MilliQ | 1.09 | 1.68 |
| 8–9 | MilliQ | 1.22 | 1.89 |
| 8 | Effluent | 2.09 | 3.24 |

An irradiation of 10 min was calculated sufficient to remove efficacy against the investigated microorganisms *P. fluorescens* and *B. subtilis* regardless of pH and water matrix. In acid milieu, $t_{act}$ is highest. The comparison between different water matrices showed that a longer irradiation time was necessary to compensate for the presence of other compounds and light absorbing substances and to remove the risk of ecotoxicity in effluent water having pH 8.

The use of UV radiation for large-scale compound removal in WWTPs would lead to rather high energy consumption. Cost of goods for sufficiently large UVA and UVC lamps may also significantly increase treatment costs. Nonetheless, UV treatment is applied for microbiological treatment of drinking water. As UV irradiation proved a rather efficient measure for pharmaceutics elimination, UV treatment could be performed on demand. This might be achieved through effluent monitoring and switching on irradiation at need.

## 4. Conclusions

The investigation of water samples from the German river Erft showed the presence of the fluoroquinolone LOM. Photoinduced removal of LOM was investigated in model water and effluent at acidic, neutral, and basic milieu und UVA and UVC irradiation using HPLC-$MS^n$ techniques. As AOP catalysts, hydrogen peroxide and titanium dioxide were used. First order kinetic models applied to concentration–time plots of LOM and its phototransformation products showed that LOM degraded fastest at pH 8 in the absence of hydrogen peroxide and titanium dioxide. Products were eliminated within 10 min of UVX irradiation. Known and novel transformation products of LOM were identified using $MS^n$. Both $MIC/IC_{50}$ determination against *P. fluorescens* and *B. subtilis* and QSAR analyses suggested the removal of ecotoxicological hazard after sufficiently long irradiation. Matrix effects of effluent or surface water required prolonged irradiation in comparison to model water. With the help of degradation rate constants and activity parameters such as MIC and irradiation time until absence of activity, the potential ecotoxicity might be predicted using Equation (4). This investigation of LOM as an example might support the search for suitable AOPs as fourth purification steps in WWTPs and the prediction of treatment time.

**Supplementary Materials:** The following are available online at http://www.mdpi.com/2571-8797/2/1/6/s1, Document 1: Supplemental Information.

**Author Contributions:** M.V., investigation; formal analysis; writing—original draft preparation; M.V., B.H., N.T., C.S. and I.B., performed the experiments; A.N.-H., supervision of biological testing; M.J., supervision, writing—original draft preparation, writing—review, and editing. All authors have read and agreed to the published version of the manuscript.

**Funding:** This research received no external funding.

**Acknowledgments:** The authors thank their institution for further financial support.

**Conflicts of Interest:** The authors declare no conflicts of interest.

## References

1. Zhang, P.; Zhou, H.; Li, K.; Zhao, X.; Liu, Q.; Li, D.; Zhao, G. Occurrence of pharmaceuticals and personal care products, and their associated environmental risks in a large shallow lake in north China. *Environ. Geochem. Health* **2018**, *40*, 1525–1539. [CrossRef]
2. Sousa, J.C.G.; Ribeiro, A.R.; Barbosa, M.O.; Pereira, M.F.R.; Silva, A.M.T. A review on environmental monitoring of water organic pollutants identified by EU guidelines. *J. Hazard. Mater.* **2018**, *344*, 146–162. [CrossRef]
3. Gogoi, A.; Mazumder, P.; Kumar, V.; Chaminda, G.G.T.; Kyoungjin, A.; Kumar, M. Groundwater for Sustainable Development Occurrence and fate of emerging contaminants in water environment: A review. *Groundw. Sustain. Dev.* **2018**, *6*, 169–180. [CrossRef]
4. Hamad, B. The antibiotics market. *Nat. Rev. Drug Discov.* **2010**, *9*, 675–676. [CrossRef] [PubMed]
5. Kümmerer, K. Antibiotics in the aquatic environment—A review—Part I. *Chemosphere* **2009**, *75*, 417–434. [CrossRef] [PubMed]
6. Nakata, H.; Kannan, K.; Jones, P.D.; Giesy, J.P. Determination of fluoroquinolone antibiotics in wastewater effluents by liquid chromatography-mass spectrometry and fluorescence detection. *Chemosphere* **2005**, *58*, 759–766. [CrossRef] [PubMed]
7. Parsons, S. *Advanced Oxidation Processes for Water and Wastewater Treatment*; IWA Publishing: London, UK, 2004.
8. Yuan, F.; Hu, C.; Hu, X.; Qu, J.; Yang, M. Degradation of selected pharmaceuticals in aqueous solution with UV and UV/H2O2. *Water Res.* **2009**, *43*, 1766–1774. [CrossRef] [PubMed]
9. Hain, E.; Wammer, K.H.; Blaney, L. Comment on Photodegradation of sulfathiazole under simulated sunlight: Kinetics, photo-induced structural rearrangement, and antimicrobial activities of photoproducts by Niu et al. Water Research 124 2017 576 e 583. *Water Res.* **2018**, *131*, 205–207. [CrossRef] [PubMed]
10. Niu, X.; Glady-croué, J.; Croué, J. Photodegradation of sulfathiazole under simulated sunlight: Kinetics, photo-induced structural rearrangement, and antimicrobial activities of photoproducts. *Water Res.* **2017**, *124*, 576–583. [CrossRef]
11. Voigt, M.; Bartels, I.; Nickisch-Hartfiel, A.; Jaeger, M. Determination of minimum inhibitory concentration and half maximal inhibitory concentration of antibiotics and their degradation products to assess the eco-toxicological potential. *Toxicol. Environ. Chem.* **2019**, *101*, 315–338. [CrossRef]
12. Dan, A.; Zhang, X.; Dai, Y.; Chen, C.; Yang, Y. Occurrence and removal of quinolone, tetracycline, and macrolide antibiotics from urban wastewater in constructed wetlands. *J. Clean. Prod.* **2020**, *252*, 119677. [CrossRef]

13. Gothwal, R. Shashidhar Occurrence of high levels of fluoroquinolones in aquatic environment due to effluent discharges from bulk drug manufacturers. *J. Hazard. Toxic Radioact. Waste* **2017**, *21*, 05016003. [CrossRef]
14. Zhang, Z.; Xie, X.; Yu, Z.; Cheng, H. Influence of chemical speciation on photochemical transformation of three fluoroquinolones (FQs) in water: Kinetics, mechanism, and toxicity of photolysis products. *Water Res.* **2019**, *148*, 19–29. [CrossRef]
15. Tammam, M.H. Photostability studies on gemifloxacin and lomefloxacin in bulk powder and dosage forms. *Eur. J. Chem.* **2014**, *5*, 73–80. [CrossRef]
16. Miranda, A.C.; Lepretti, M.; Rizzo, L.; Caputo, I.; Vaiano, V.; Sacco, O.; Lopes, W.S.; Sannino, D. Surface water disinfection by chlorination and advanced oxidation processes: Inactivation of an antibiotic resistant E. coli strain and cytotoxicity evaluation. *Sci. Total Environ.* **2016**, *554–555*, 1–6. [CrossRef]
17. Guo, C.; Wang, K.; Hou, S.; Wan, L.; Lv, J.; Zhang, Y.; Qu, X.; Chen, S.; Xu, J. $H_2O_2$and/or $TiO_2$photocatalysis under UV irradiation for the removal of antibiotic resistant bacteria and their antibiotic resistance genes. *J. Hazard. Mater.* **2017**, *323*, 710–718. [CrossRef]
18. Vracko, M. *Mathematical (Structural) Descriptors in QSAR: Applications in Drug Design and Environmental Toxicology*; Elsevier: Amsterdam, The Netherlands, 2016; Volume 1, ISBN 9781681081977.
19. Zhu, H.; Shen, Z.; Tang, Q.; Ji, W.; Jia, L. Degradation mechanism study of organic pollutants in ozonation process by QSAR analysis. *Chem. Eng. J.* **2014**, *255*, 431–436. [CrossRef]
20. Kruhlak, N.L.; Contrera, J.F.; Benz, R.D.; Matthews, E.J. Progress in QSAR toxicity screening of pharmaceutical impurities and other FDA regulated products. *Adv. Drug Deliv. Rev.* **2007**, *59*, 43–55. [CrossRef]
21. Abdel-Aziz, A.A.M.; Asiri, Y.A.; Al-Agamy, M.H.M. Design, synthesis and antibacterial activity of fluoroquinolones containing bulky arenesulfonyl fragment: 2D-QSAR and docking study. *Eur. J. Med. Chem.* **2011**, *46*, 5487–5497. [CrossRef]
22. Anquetin, G.; Greiner, J.; Mahmoudi, N.; Santillana-Hayat, M.; Gozalbes, R.; Farhati, K.; Derouin, F.; Aubry, A.; Cambau, E.; Vierling, P. Design, synthesis and activity against Toxoplasma gondii, Plasmodium spp., and Mycobacterium tuberculosis of new 6-fluoroquinolones. *Eur. J. Med. Chem.* **2006**, *41*, 1478–1493. [CrossRef]
23. Toolaram, A.P.; Haddad, T.; Leder, C.; Kümmerer, K. Initial hazard screening for genotoxicity of photo-transformation products of ciprofloxacin by applying a combination of experimental and in-silico testing. *Environ. Pollut.* **2016**, *211*, 148–156. [CrossRef]
24. Voigt, M.; Bartels, I.; Nickisch-Hartfiel, A.; Jaeger, M. Elimination of macrolides in water bodies using photochemical oxidation. *AIMS Environ. Sci.* **2018**, *5*, 372–388. [CrossRef]
25. Voigt, M.; Bartels, I.; Nickisch-Hartfiel, A.; Jaeger, M. Photoinduced degradation of sulfonamides, kinetic, and structural characterization of transformation products and assessment of environmental toxicity. *Toxicol. Environ. Chem.* **2017**, *99*, 1304–1327. [CrossRef]
26. Voigt, M.; Jaeger, M. On the photodegradation of azithromycin, erythromycin and tylosin and their transformation products—A kinetic study. *Sustain. Chem. Pharm.* **2017**, *5*, 131–140. [CrossRef]
27. Kuhn, H.; Braslavsky, S.E.; Schmidt, R. Chemical Actinometry. *IUPAC Tech. Rep.* **2004**, *76*, 2105–2146. [CrossRef]
28. Hatchard, C.G.; Parker, C. A New Sensitive Chemical Actinometer. II. Potassium Ferrioxalate as a Standard Chemical Actinometer. *Proc. R. Soc. A Math. Phys. Eng. Sci.* **1956**, *235*, 518–536.
29. Sun, L.; Bolton, J.R. Determination of the Quantum Yield for the Photochemical Generation of Hydroxyl Radicals in $TiO_2$ Suspensions. *J. Phys. Chem.* **1996**, *100*, 4127–4134. [CrossRef]
30. Kochany, J.; Bolton, J.R. Mechanism of photodegradation of aqueous organic pollutants. 2. Measurement of the primary rate constants for reaction of hydroxyl radicals with benzene and some halobenzenes using an EPR spin-trapping method following the photolysis of hydrogen peroxide. *Environ. Sci. Technol.* **1992**, *26*, 262–265. [CrossRef]
31. Kochany, J.; Bolton, J.R. Of aqueous organic pollutants. 1. EPR spin-trapping technique for the determination of hydroxyl radical rate constants in the photooxidation of chlorophenols following. *J. Phys. Chem.* **1991**, *95*, 5116–5120. [CrossRef]
32. Chan, P.Y.; Gamal El-Din, M.; Bolton, J.R. A solar-driven UV/Chlorine advanced oxidation process. *Water Res.* **2012**, *46*, 5672–5682. [CrossRef]
33. Li, K.; Zhang, P.; Ge, L.; Ren, H.; Yu, C.; Chen, X.; Zhao, Y. Concentration-dependent photodegradation kinetics and hydroxyl-radical oxidation of phenicol antibiotics. *Chemosphere* **2014**, *111*, 278–282. [CrossRef] [PubMed]

34. Fernández, R.L.; McDonald, J.A.; Khan, S.J.; Le-Clech, P. Removal of pharmaceuticals and endocrine disrupting chemicals by a submerged membrane photocatalysis reactor (MPR). *Sep. Purif. Technol.* **2014**, *127*, 131–139. [CrossRef]
35. Jung, Y.J.; Kim, W.G.; Yoon, Y.; Kang, J.-W.; Hong, Y.M.; Kim, H.W. Removal of amoxicillin by UV and UV/H2O2 processes. *Sci. Total Environ.* **2012**, *420*, 160–167. [CrossRef] [PubMed]
36. Voigt, M.; Savelsberg, C.; Jaeger, M. Photodegradation of the antibiotic spiramycin studied by high-performance liquid chromatography-electrospray ionization-quadrupole time-of-flight mass spectrometry. *Toxicol. Environ. Chem.* **2017**, *99*, 624–640. [CrossRef]
37. Mauser, H. *Formale Kinetik*; Experimentelle Methoden der Physik und der Chemie; Bertelsmann-Universitätsverlag: Düsseldorf, Germany, 1974; ISBN 9783571191889.
38. Mazellier, P.; Méité, L.; De Laat, J. Photodegradation of the steroid hormones 17beta-estradiol (E2) and 17alpha-ethinylestradiol (EE2) in dilute aqueous solution. *Chemosphere* **2008**, *73*, 1216–1223. [CrossRef] [PubMed]
39. Wiegand, I.; Hilpert, K.; Hancock, R.E.W. Agar and broth dilution methods to determine the minimal inhibitory concentration (MIC) of antimicrobial substances. *Nat. Protoc.* **2008**, *3*, 163–175. [CrossRef]
40. EUCAST. *ISO 20776-1:2007*; IOS: Geneva, Switzerland, 2007.
41. Martin, T.; Harten, P.; Venkatapathy, R.; Young, D. *TEST (Toxicity Estimation Software Tool) Ver 4.1.*; U.S. Environmental Protection Agency: Washington, DC, USA, 2016.
42. QSAR Toolbox. *Laboratory of Mathematical Chemistry*; University Bourgas: Bourgas, Bulgaria, 2012.
43. Gao, L.; Shi, Y.; Li, W.; Niu, H.; Liu, J.; Cai, Y. Occurrence of antibiotics in eight sewage treatment plants in Beijing, China. *Chemosphere* **2012**, *86*, 665–671. [CrossRef]
44. Li, W.; Shi, Y.; Gao, L.; Liu, J.; Cai, Y. Occurrence, distribution and potential affecting factors of antibiotics in sewage sludge of wastewater treatment plants in China. *Sci. Total Environ.* **2013**, *445–446*, 306–313. [CrossRef]
45. Rosa, J.M.; Tambourgi, E.B.; Vanalle, R.M.; Carbajal Gamarra, F.M.; Curvelo Santana, J.C.; Araújo, M.C. Application of continuous H2O2/UV advanced oxidative process as an option to reduce the consumption of inputs, costs and environmental impacts of textile effluents. *J. Clean. Prod.* **2019**, *246*, 119012. [CrossRef]
46. Al-Mamun, M.R.; Kader, S.; Islam, M.S.; Khan, M.Z.H. Photocatalytic activity improvement and application of UV-$TiO_2$ photocatalysis in textile wastewater treatment: A review. *J. Environ. Chem. Eng.* **2019**, *7*, 103248. [CrossRef]
47. Oppenländer, T. *Photochemical Purification of Water and Air: Advanced Oxidation Processes (AOPs): Principles, Reaction Mechanisms, Reactor Concepts (Chemistry)*; WILEY-VCH Verlag: Weinheim, Germany, 2003; ISBN 3-527-30463-7.
48. Homem, V.; Santos, L. Degradation and removal methods of antibiotics from aqueous matrices–a review. *J. Environ. Manag.* **2011**, *92*, 2304–2347. [CrossRef] [PubMed]
49. Calza, P.; Minella, M.; Demarchis, L.; Sordello, F.; Minero, C. Photocatalytic rate dependence on light absorption properties of different $TiO_2$ specimens. *Catal. Today* **2020**, *340*, 12–18. [CrossRef]
50. Fasani, E.; Mella, M.; Caccia, D.; Tassi, S.; Fagnoni, M.; Albini, A. The photochemistry of lomefloxacin. An aromatic carbene as the key intermediate in photodecomposition. *Chem. Commun.* **1997**, 1329–1330. [CrossRef]
51. Budai, M.; Gróf, P.; Zimmer, A.; Pápai, K.; Klebovich, I.; Ludányi, K. UV light induced photodegradation of liposome encapsulated fluoroquinolones: An MS study. *J. Photochem. Photobiol. A Chem.* **2008**, *198*, 268–273. [CrossRef]
52. Fasani, E.; Albini, A.; Mella, M.; Rampi, M.; Negra, F.B. Light and drugs: The photochemistry of fluoroquinolone antibiotics. *Int. J. Photoenergy* **1999**, *1*, 7–11. [CrossRef]
53. Fasani, E.; Monti, S.; Manet, I.; Tilocca, F.; Pretali, L.; Mella, M.; Albini, A. Inter- and intramolecular photochemical reactions of fleroxacin. *Org. Lett.* **2009**, *11*, 1875–1878. [CrossRef]
54. Liu, C.; Nanaboina, V.; Korshin, G.V.; Jiang, W. Spectroscopic study of degradation products of ciprofloxacin, norfloxacin and lomefloxacin formed in ozonated wastewater. *Water Res.* **2012**, *46*, 5235–5246. [CrossRef]

*clean technologies*

MDPI

*Article*

# Stimulating Nitrogen Biokinetics with the Addition of Hydrogen Peroxide to Secondary Effluent Biofiltration

**Liron Friedman [1,2,3], Hadas Mamane [2], Kartik Chandran [3], Martin Jekel [4], Haim Cikurel [5], Uwe Hübner [6], Michael Elgart [7], Shlomi Dagan [7], Jorge Santo-Domingo [8] and Dror Avisar [1,*]**

1 The Water Research Center, School of Earth Sciences, Faculty of Exact Sciences, Tel Aviv University, Tel Aviv 69978, Israel; lirfri@gmail.com
2 School of Mechanical Engineering, Faculty of Engineering, Tel Aviv University, Tel Aviv 69978, Israel; hadasmg@tauex.tau.ac.il
3 Department of Earth and Environmental Engineering, Columbia University, New York, NY 4711, USA; kc2288@columbia.edu
4 Department of Water Quality Control, Technical University of Berlin, 10623 Berlin, Germany; martin.jekel@tu-berlin.de
5 Environmental Services, consultant, Bat-Yam 59555, Israel; chikurel@netvision.net.il
6 Urban Water Systems Engineering, Technical University of Munich, 85748 Garching, Germany; u.huebner@tum.de
7 Weizmann Institute of Science, Molecular cell biology, Rehovot 76100, Israel; Elgart@gmail.com (M.E.); Shlomi.dagi@gmail.com (S.D.)
8 US Environmental Protection Agency; National Risk Management Research Laboratory, Cincinnati, OH 45268, USA; santodomingo.jorge@epa.gov
* Correspondence: droravi@tauex.tau.ac.il; Tel.: +972-3-6405534

Received: 22 December 2019; Accepted: 20 January 2020; Published: 1 February 2020

**Abstract:** Tertiary wastewater treatment could provide a reliable source of water for reuse. Amongst these types of wastewater treatment, deep-bed filtration of secondary effluents can effectively remove particles and organic matter; however, $NH_4^+$ and $NO_2^-$ are not easily removed. This study examined the feasibility of stimulating microbial activity using hydrogen peroxide ($H_2O_2$) as a bio-specific and clean oxygen source that leaves no residuals in the water and is advantageous upon aeration due to the solubility limitations of the oxygen. The performance of a pilot bio-filtration system at a filtration velocity of 5–6 m/h, was enhanced by the addition of $H_2O_2$ for particle, organic matter, $NH_4^+$, and $NO_2^-$ removal. Hydrogen peroxide provided the oxygen demand for full nitrification. As a result, influent concentrations of 4.2 ± 2.5 mg/L N-$NH_4^+$ and 0.65 ± 0.4 mg/L N-$NO_2$ were removed during the short hydraulic residence time (HRT). In comparison, filtration without $H_2O_2$ addition only removed up to 0.6 mg/L N-$NH_4^+$ and almost no N-$NO_2^-$. A DNA metagenome analysis of the functional genes of the media biomass reflected a significant potential for simultaneous nitrification and denitrification activity. It is hypothesized that the low biodegradability of the organic carbon and $H_2O_2$ addition stimulated oxygen utilization in favor of nitrification, followed by the enhancement of anoxic activity.

**Keywords:** hydrogen peroxide; high-rate biofiltration; nitrification; denitrification

## 1. Introduction

The tertiary and advanced treatment of wastewater improves water quality, enabling us to meet environmental concerns but also find a potential source of water for reuse [1,2]. The filtration of coagulated–flocculated effluent via high-rate deep-bed filters (velocities between 5 and 20 m/h) is widely used for tertiary wastewater treatment [3]. The removal mostly combines particles and some dissolved organic matter (15%–20%) [4]. A hybrid process of filtration and bioactivity, termed biofiltration, allows significant nutrient and organic matter removal in addition to particle filtration [5–7].

Molecular methods provide better insight into the processes that are essential for optimal system design. Recent research on the treatment of secondary effluent or groundwater, as sand-filtration and AOP processes, such as UV/$H_2O_2$, made use of advanced methods as metagenomics, metaproteomics or metatranscriptomics, to analyze the structure of the community attached to the media and determine the related metabolism [8–12].

In biological wastewater systems, dissolved oxygen (DO) is a key parameter in system performance and plays an important role in biotransformation pathways and rates [13,14]. The competition for oxygen among ammonia-oxidizing bacteria (AOB), nitrite-oxidizing bacteria (NOB), and ordinary heterotrophic organisms (OHO) results in oxygen depletion; this, in turn, increases the potential for anoxic (carbon oxidation through reduction of $NO_2^-$ or $NO_3^-$) pathways, such as full or partial denitrification, depending on substrate availability and biodegradability [15–18]. Compared to suspended growth systems, nitrification activity can increase significantly in attached biomass systems when oxygen is more available [19]. DO concentration in the attached biomass is limited by the temperature of the solute (~8 mg/L at 25 °C) and decreases along the biofilm depth, resulting in increased denitrification rates [20,21]. To overcome these limitations, a few studies have examined the use of 100% pure oxygen or hydrogen peroxide (hereafter $H_2O_2$) degradation (Equation (1)) in attached-biomass systems [22–25].

$$1H_2O_2 \underset{Catalase}{\Rightarrow} 0.53H_2O + 0.47O_2 \left(K_{cat} = 10^7\,\text{sec}^{-1}\right) \quad (1)$$

$H_2O_2$ degradation by the enzyme catalase. Stochiometry is by weight (g); $K_{cat}$—catalytic reaction rate [26].

Catalase is a significant component of the cell defense mechanism against oxidative stress, it scavenges two molecules of $H_2O_2$ into water and molecular oxygen in one of the highest documented enzyme reaction rates [27,28].

$H_2O_2$ has advantages as a supplemental oxygen source, namely, high solubility in aqueous solution and an improved mass transfer of dissolved $H_2O_2$ (liquid–liquid) into the permeable biofilm compared to oxygen that depends also on the oxygen–water transfer into the liquid phase. In addition, rapid $H_2O_2$ degradation is bacterium-specific and leaves no persistent residual traces in the water. The disinfection characteristics of $H_2O_2$ could limit microbial-based reactors, which depend on concentration and time (C × t). Significantly tolerant degradation rates and oxygen utilization have been observed in biofilms, in comparison to suspended bacteria, in addition to changes in microbial community structure [29–33]. However, the application of $H_2O_2$ as a supplemental oxygen source via high-rate media filtration for $NH_4^+$ removal at a short hydraulic residence time (HRT) has never been examined. In this study, we demonstrate the impact of adding $H_2O_2$ as a supplemental oxygen source on the operational efficiency of a secondary effluent filtration system with biologically active media. The setup design was aimed at combining the removal of particles, $NH_4^+$, and, especially, $NO_2^-$ to reduce ozone demand for a subsequent ozonation treatment (detailed in Zucker et al., [34]). In short, the biofilter was part of a multistep pretreatment unit prior to ozonation to allow better soil aquifer treatment (SAT) performance by reducing particle clogging and minimizing oxygen demand during infiltration in the upper-layer SAT. The specific goals of this study were to examine whether significant biotransformation can be achieved with high-rate (relative to standard biofiltration) filtration conditions and $H_2O_2$ addition.

## 2. Experiments

### 2.1. Pretreatment System Setup

The biofilter system was operated in direct-filtration at the pilot site of the Shafdan wastewater treatment plant (WWTP), operated by Mekorot. The biofilter system, detailed in Figure 1, was part of the multistage setup specified herein:

- Feed flow was 6 $m^3$/h (144 $m^3$/day) of the secondary effluent from the WWTP (Figure 1a) following a 500-μm wire-mesh filtration (Figure 1b) to remove coarse particles.
- Coagulation/flocculation was carried out by injecting polyaluminum chloride (PACl, 18% $Al_2O_3$) by peristaltic pump (Figure 1c) to achieve a final concentration of ~2.7–3.6 mg/L as PACl. Flocculation was performed in a modified flocculator, which consisted of a pressure filter (Figure 1d) with ~15 min hydraulic retention time (HRT). PACl was chosen for this study as it has been shown to be preferable for flocculation and is widely used [35,36].
- Following flocculation, $H_2O_2$ was added (10%) to the inlet of the high-rate biofilter (Figure 1f) using a peristaltic pump (Figure 1e). The final concentration of 27 mg/L was achieved to provide a surplus of DO for full nitrification of 3.5 mg/L-N $NH_4^+$ and 0.5 mg/L-N $NO_2^-$. The biofilter tank had a surface area of 1.13 $m^2$, 1.2 m diameter, 1.1 $m^3$ media volume, and 36% free headspace (additional characteristics are specified in Zucker et al., [34]).

During the adjustment period with stepwise $H_2O_2$ addition, an average measured concentration of 4.29 ± 0.56 mg/L DO at the outlet of the filter was obtained with the addition of ~27 mg/L $H_2O_2$. The theoretical calculation of oxygen mass balance is provided in Appendix A.1. During the research period, the loads of $NH_4^+$ in the secondary effluent increased due to changes in the process of the plant, which enabled investigating the system performance for $NH_4^+$ oxidation and N removal under higher $NH_4^+$ loads and $O_2$ limitation.

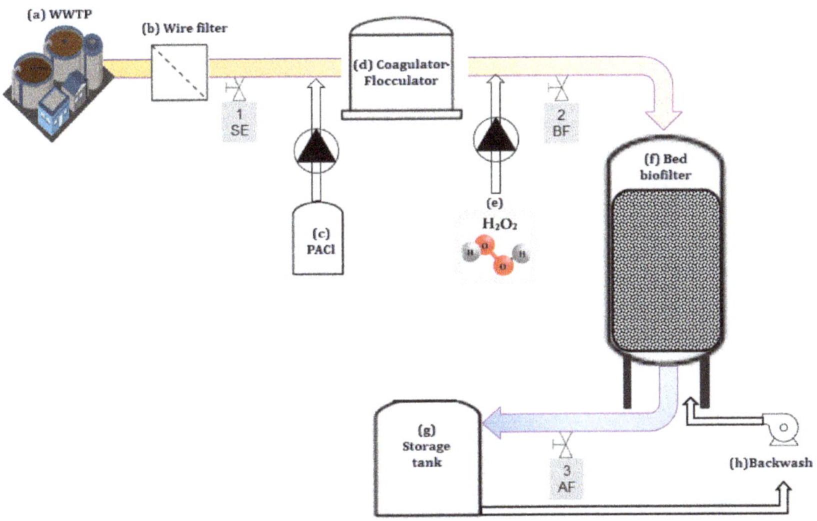


**Figure 1.** Diagram of biofiltration setup. (**a**) Secondary effluent from the Shafdan wastewater Table 500 m, (**b**) wire filter, (**c**) the addition of polyaluminum chloride (PACl) from a tank, (**d**) the coagulator–flocculator tank, (**e**) the addition of $H_2O_2$, (**f**) the bed biofilter, (**g**) the storage tank, and (**h**) the backwash pump. Sampling points were located (1) right after the wire-mesh filter, termed the secondary effluent (SE), (2) after flocculation–coagulation and $H_2O_2$ addition, but before the biofilter (BF) and (3) after the biofilter (AF).

### 2.2. *System Specification and Operational Parameters*

The biofiltration was operated with a modified active media filter, operated in downflow mode, with a filtration velocity of 5–6 m/h (~5 min hydraulic residence time (HRT, the low range of high-rate filtration) and a backwash cycle every 12 h. The filtered effluent water was stored in a 10-$m^3$ tank (Figure 1g) and was used for backwash. Adjustments were performed to avoid the clogging of the filter while maintaining a steady and active biomass to avoid the detachment of biofilm from the media

(Appendix A.2). The microbial community of the biomass was indigenous and developed over time by feeding the biofilter with the secondary effluent. It was assumed that the accumulation and growth of viable bacteria in the filter, most of which were fixed to the media, was the basis for the bacterial activity [6,37–39].

### 2.3. Solute Sampling and Analyses

After two months of system operation without $H_2O_2$ addition, samples were further taken over for a period of four months with stable performance. Stabilization after $H_2O_2$ addition took also about two months and then a sampling campaign was conducted every 2–4 weeks for over two years. At every sampling campaign, samples were collected 6–7 h after the start of the morning filtration cycle (middle of the cycle) from sampling faucet located past the mesh filter, named the secondary effluent (SE), before the biofilter (BF) and after the biofilter (AF). An analysis of water characteristics was conducted at the Shafdan WWTP laboratory specified in the Appendix A.3.

### 2.4. Particle Sampling and Analysis

Effluent particles and aggregates were analyzed by Micro Flow Imaging (MFI) technology (DPA 4100, Protein Simple Inc., Ottawa, ON, Canada). This apparatus employs a digital camera with an illumination and magnification system to capture in-situ images of suspended particles in a flowing sample. A detailed description of this analysis is published elsewhere [32,36]. An analysis was conducted on particles of between ~2 and 400 μm.

### 2.5. Media Sampling and Metagenome Analysis

A metagenome analysis was conducted to investigate the functional potential of the microbial community in the media. In the middle of the operation cycle (6 h), the biofilter was emptied and a sample of 50 g anthracite media was taken from a 40-cm depth, thoroughly mixed for homogenization, and immediately stored in dry ice. Sampling was conducted a few days after the sampling campaign designated in star in Figure A1 in the Appendix A. DNA from the media was extracted using an Exgene Soil DNA Kit. The whole-genome (shotgun) libraries were prepared using the TruSeq DNA PCR-Free HT Library Preparation Kit (Illumina, San Diego, CA, USA) and were sequenced using an Illumina MiSeq benchtop sequencer with a target fragment length of 250 bp. After the removal of adapters and low-quality reads, ~8.2 M reads were left with an average length of 225 base pairs. (Appendix A Figure A2). High-quality reads were mapped to N-cycle related genes using BWA [40]. The potential for dominant biotransformation of the microbial community was deduced from the relative abundance of specific functional genes. The genes for nitrification (amoA, nxrB, cyt c), denitrification (narG, nirK, nirS, norB, nosZ), and other various metabolic pathways, such as methane (pmoA), hydrogen sulfide (drsEFH), and hydrogen (HiFe) oxidation were analyzed and compared on a log scale [9]. The gene catalog was extracted from NCBI's nucleotide database. Genes aligning to the reads were sorted according to their respective identified organisms, which varied between 6%–19% among all the reads related to each gene (Figure A2 in the Appendix A).

## 3. Results

The goal of this study was to design a biofilter with short HRT to support nitrification activity, by surplus oxygen, for the removal of $NO_2^-$ and $NH_4^+$. The system showed a significant removal of $NO_2^-$ and $NH_4^+$ when $H_2O_2$ was added in comparison to the control. The biofilter was optimized in the matter of $H_2O_2$ addition and evaluated as a single step in an integrated treatment with the ultimate goal to produce a streamwith lower oxygen demand for following SAT or non-potable reuse purposes.

### 3.1. Particle Distribution

To determine biofilter performance in removing particles as a function of equivalent circular diameter (ECD), and the particles' size and the size distribution in the secondary effluent, a dynamic image analysis was made. The particle-size distribution (PSD) of secondary effluents based on 13 campaigns is presented in Figure A1 in the Appendix A, taken at the SE sampling point. In general, the PSD was similar in comparison to the high variation in total particle number. Almost all (99%) of the particles were analyzed (<50 μm), with the highest variation in the fraction with ECD = 2–3 μm. The total particle removal performance was found to be highly efficient (95 ± 2%) for all sampling campaigns. No difference was observed in particle removal without $H_2O_2$. A typical PSD analysis of the secondary effluent after 500 μm, post flocculation and after the biofilter, is presented in Figure A1 in the Appendix A. Flocculation before biofiltration increased particle counts due to the flocculation of the macromolecules and colloid particles (<2 μm), which were not analyzed. As expected, particle concentration decreased dramatically after biofiltration for all ECDs.

### 3.2. $H_2O_2$ Decomposition

The concentration of DO before filtration varied between 3–5 mg/L. No $H_2O_2$ was detected at the inlet or outlet during an operation without $H_2O_2$ addition, while the DO values at the outlet were <1 mg/L, which clearly indicated maximum oxygen utilization and reflected anoxic conditions [20]. The preliminary lab-bench experiments revealed no decomposition of $H_2O_2$ when mixed with effluents within the interval of the HRT (20 min), as also documented by Lakretz et al. [32]. When $H_2O_2$ was added, the measured $H_2O_2$ concentration at the inlet was 27 ± 2 mg/L, and <1 mg/L at the outlet, which affirms full $H_2O_2$ degradation. Furthermore, DO values after filtration were similar or slightly higher than before filtration. Thus, we conclude that $H_2O_2$ decomposition in the system occurs almost solely via bacterial activity and functions as a source of available oxygen.

### 3.3. Reference Measurements—No Addition of $H_2O_2$

Sampling campaigns were conducted without the addition of $H_2O_2$ as a reference: DO values at the outlet were <1 mg/L, which reflects maximum oxygen utilization and anaerobic conditions inside the reactor [20]. The dissolved organic carbon (DOC)- and chemical oxygen demand (COD)-removal rates were similar (13% ± 0.3%, and 11% ± 2%, respectively) and were mainly caused by flocculation/filtration. Data were insufficient for the ultra violet absorption (UVA) measurements. The average $NH_4^+$, $NO_2^-$ and $NO_3^-$ concentrations at the BF sampling point were 3.1 ± 0.60 mg/L-N, 0.37 ± 0.07 mg/L-N, and 0.34 ± 0.05 mg/L-N, respectively. While $NH_4^+$ removal was constant at 0.6 ± 0.08 mg/L-N, $NO_2^-$, removal was not constant, and in some samples no removal of $NO_2^-$ was observed. The average increase in $NO_3^-$ concentration was 0.73 ± 0.21 mg/L-N. The calculated nitrogen loss was fairly low and did not exceed 4 ± 1%, which suggests removal by filtration rather than via denitrification.

### 3.4. Organic Carbon Removal with $H_2O_2$ Addition

The biofilter removal rates of DOC and UVA were stable at 22% ± 5% and 20% ± 3%, respectively, regardless of the variation in DOC (10.4 ± 1.4 mg/L) and UVA (217 ± 13 $m^{-1}$) at the inlet. Without $H_2O_2$, the ΔDOC, $\Delta UVA_{254}$ and ΔCOD was 1.6 ± 0.4 mg/L, 7 ± 1 $m^{-1}$ and 3.3 ± 1 mg/L, respectively. With addition of $H_2O_2$, the ΔDOC, $\Delta UVA_{254}$ and ΔCOD increased to 2.3 ± 0.7 mg/L, 44 ± 7 1/m and 14.6 ± 4 mg/L, respectively, indicating additional organic carbon utilization by bacteria (Appendix A Figure A5: concentrations and the removal of DOC and $UVA_{254}$ over sampling campaigns before the biofilter and after the biofilter). The oxygen utilization, in favor of nitrification rather than carbon oxidation, reflects the low biodegradability of organic matter, which is typical for effluent post-secondary treatment with flocculation–filtration removal [41–44]. This is also supported by no removal of most trace organic compounds in the biofilter, besides Acesulfame and Iopromide that were removed by 60% and 30%, respectively [34]. These results show that the biofilter was efficient and robust at removing particles, DOC, and UVA.

### 3.5. $NH_4^+$ and $NO_3^-$ Transformation, Variation, and Nitrogen Mass Balance

To better illustrate the potential nitrogen tranformations in the biofilter due to $H_2O_2$ addition, and especially the $NH_4^+$ removal (mg/L-N) ability of the system, the parameters at all the figures were sorted by $NH_4^+$ concentrations at the inlet (BF) sampling point, as illustrated in Figure 2. It can be clearly observed that the concentration of $NO_3^-$ that increased at the outlet of the biofilter is in parallel to the decrease in $NH_4^+$ concentration (Figure 2). Full $NH_4^+$ removal with $H_2O_2$ addition was obtained when the BF concentration was lower than 4 mg/L-N, whereas above 4 mg/L-N residual $NH_4^+$ was detected at the AF sampling point. The average $NO_2^-$ concentration of 0.64 ± 0.4 mg/L-N was fully removed after the biofilter (AF) for all samples (Supporting data Figure A6: removal of $NO_2^-$), indicating rapid $NO_2^-$ oxidation to $NO_3^-$ (nitratation). Thus, $H_2O_2$ addition enabled a significantly higher nitrification rate. Given that the nitrification rate in the biofilter was stable and was solely dependent on $H_2O_2$ addition, regardless of particle numbers and removal, it is suggested that most of the microbial activity occurred in the biomass attached to the media, perhaps independent of the filtration cake and its size. The ammonium oxidation rate (AOR) (the ratio of formed $NO_3^-$ to the transformed $NH_4^+$ and $NO_2^-$), is calculated according to Equation (2) and shown for all sampling points

$$\mathrm{AOR} = -\frac{\Delta[\mathrm{NO_3^-} - \mathrm{N}]}{\left\{\Delta\left[\mathrm{NH_4^+} - \mathrm{N}\right] + \Delta[\mathrm{NO_2^-} - \mathrm{N}]\right\}} \tag{2}$$

ONT = Ammonification, organic nitrogen transformed to $NH_4^+$
Nit = Nitrification
NX = Anoxic/AMX (anaerobic ammonia oxidation) nitrogen transformation to dinitrogen
Inorganic nitrogen balance rate to indicate dominant pathway:

When AOR > 100%, ONT + Nit >> NX
When AOR < 100%, significant NX
When AOR ≈ 100%, Nit >> ONT, NX

In most samples, AOR revealed values of less than 100% (Figure 2, green dashed line), indicating the loss of ammoniacal nitrogen as dinitrogen, which will be further detailed.

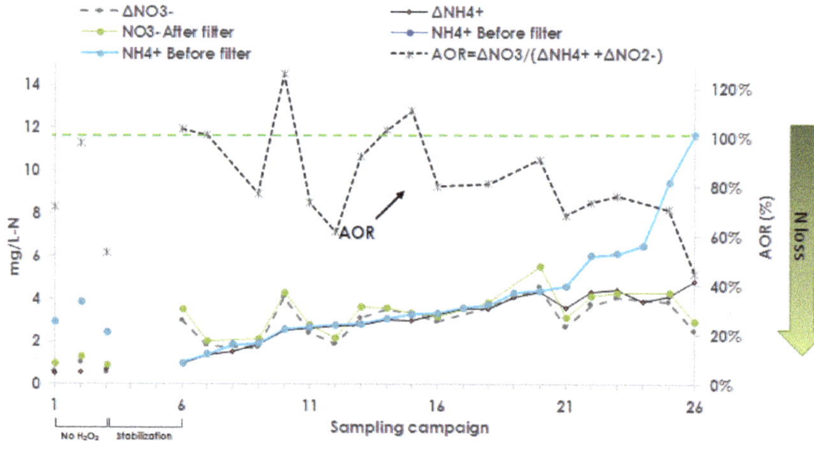


**Figure 2.** Concentrations of $NH_4^+$ before and after the filter, $NO_3^-$ after the filter, and the delta of $NH_4^+$ and $NO_3^-$ (mg/L-N) are presented. The gray dashed line represents the $NH_4^+$ oxidation rate, which is the ratio between the introduced $NO_3^-$ and the sum of the removed $NH_4^+$ and $NO_2^-$ (%). The green dashed line represents the complete (100%) calculated mass balance between removed $NH_4^+$ and $NO_2^-$ and introduced $NO_3^-$. The sampling campaigns were sorted by feed values of $NH_4^+$. The first three values represent the control without $H_2O_2$ addition.

Moreover, the calculated AOR was less than ~70% when $NH_4^+$ values at the BF sampling point were higher than 4 mg/L-N. Values over 100% can be explained by the contribution of $NH_4^+$ via a higher rate of organic nitrogen (amino acids) decomposition than nitritation. The relatively low removal rates of organic compounds by the biofilter when $H_2O_2$ was added, in comparison to $NH_4^+$ and $NO_2^-$, suggests that the level of organic carbon degradability was a limiting factor in favoring oxygen utilization for nitrification on the one hand and carbon utilization for denitrification on the other.

### 3.6. Mass Balance

Calculations of the total nitrogen mass balance are presented in Figure 3, considering organic nitrogen values, revealed an average nitrogen loss of 2.24 ± 1 mg-N/L with $H_2O_2$ and 0.25 ± 0.06 mg-N/L at the control. In addition, a higher nitrogen loss was calculated in parallel to the increase in $NH_4^+$ concentration at the inlet. Calculations to evaluate the nitrogen loss in favor of assimilation with $H_2O_2$ addition revealed that the assimilation rate was around 50% from all nitrogen removed (Table A1). This indicates that significant nitrogen loss was due to anoxic activity in the biofilter, which occurred simultaneously with nitrification. In addition, denitrification credit, the calculated oxygen that was credited due to utilization of $NO_3^-$ as electron acceptor, was significant and varied between 1–7 mg/L of COD and also increased in parallel to the increase in $NH_4^+$ concentrations at the inlet. To explain the higher nitrogen loss under aerobic conditions that was enhanced by the oxygen surpluss as $H_2O_2$, we hypothesized that $H_2O_2$, due to a lack of solubility limitations, better penetrates into the depth of the biofilm, decomposes, and stimulates activity.

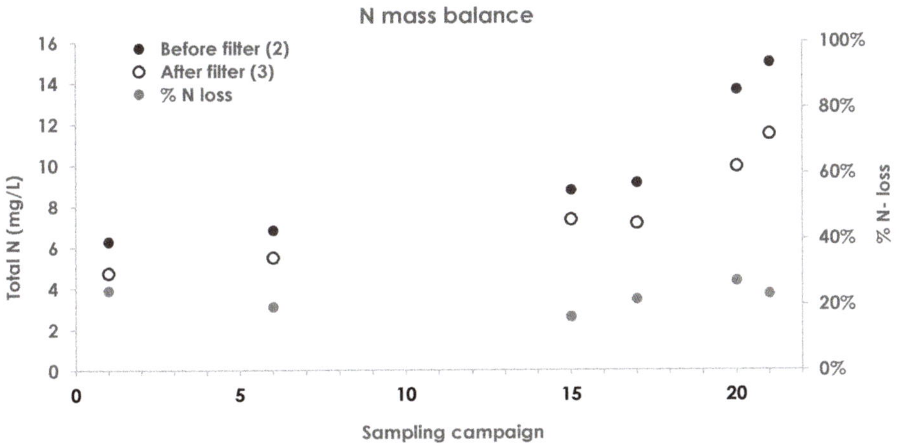


**Figure 3.** Accumulated inorganic and organic nitrogen concentrations (mg/L-N) before (BF) and after (AF) the biofilter and removal rates (%).

### 3.7. Functional Potential of the Biomass

The Metagenomics preformed on the sample yielded 8.2 M high-quality reads, out of which, 3.6 M were aligned to ~1250 organisms, when >98% are bacteria, both in the abundance and number of strains. The resulting outcome indicated a highly diverse bacterial community, with a calculated Shannon diversity index value of 5.6 as opposed to the theoretical value of 7.13 for a sample with equal diversity. The relative abundance of functional genes in the biofilter biomass is presented in Figure 4. The relative abundance of genes involved in nitrification (AmoA and NxrB) was higher, by 3 to 4 orders of magnitude, than the genes involved in other electron donor pathways (as drsEFH or aprA). In addition, the log relative abundance of the genes involved in denitrification and oxygen utilization (cyt c oxidases) was also significantly higher in comparison to the other various metabolic pathways examined.

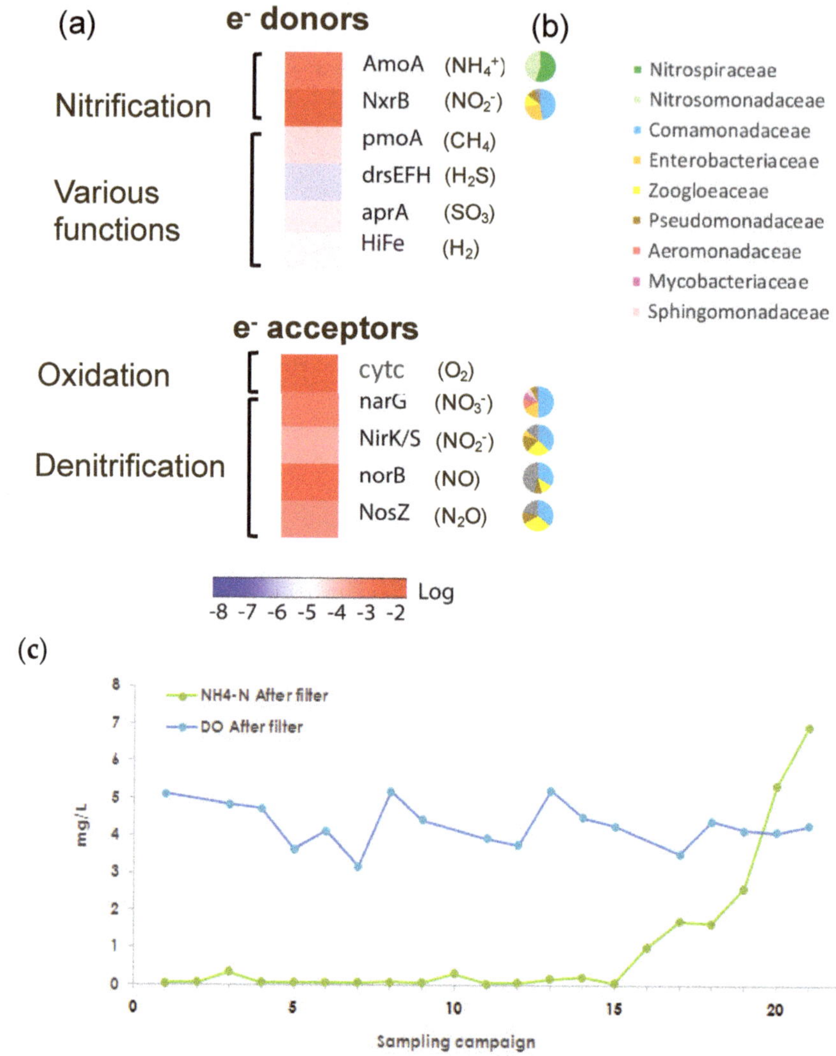


**Figure 4.** Log-transformed relative abundances of the marker genes of enzymes involved in nitrification, denitrification, and other metabolic pathways in the biofilter while $H_2O_2$ was added with various electron donors and acceptors (**a**). For each functional gene, the composition of classified genera is presented (**b**). Concentrations of ammonia and oxygen in the system outlet (**c**).

Considering that cyt c is harbored by AOB (nitritation) [45], as well as denitrifying organisms (denitrification) [46], the relative high abundance of this gene is consistent with the experimental results of simultaneous nitrification and denitrification activity (Figure 2). The higher relative abundance obtained for the gene of $NO_2^-$ oxidation (nxrB) over $NH_4^+$ oxidation, might suggest the possible occurrence of a "ping pong" mechanism of cyclic $NO_2^-$ oxidation and reduction, resulting in a higher $NO_2^-$ oxidation rate than $NH_4^+$ [47]. In parallel, the composition of identified genera of each of the N related genes (Figure 4b) revealed a variety of heterotrophs capable of conducting denitrification, when the genera belonging to the family *Comamondaceae* was the most dominant one. Interestingly, the identified genus related to $NH_4^+$ oxidation was related to AOB and NOB, while the identified genus related to $NO_2^-$ oxidation was of heterotrophs. This might suggest that the involvement of commamox and heterotrophs in nitrification [48–50].

The comparison of the functional potential of the microbial community reflects clear dominance for N biotransformation by simultaneous nitrification and denitrification. The significant anoxic potential reflected from the relative abundance of functional genes can be explained by the denitrifying community capable of being active at the presence of oxygen (low half-saturation rate of oxygen inhibition, $K_{OI}$ = 0.1–0.2 mg/L) [43].

Moreover, favoring oxygen utilization for nitrification, with a C/N ratio similar to wastewater, can be explained by the selection of organisms that are more competitive and characterized with a low substrate half-saturation rate coefficient ($K_{ONitrospira}$ = 0.1–0.2 mg/L, $K_{OAOB}$ = 0.26–0.15 mg/L), which is more typical to nitrifiers in comparison to heterotrophs [51,52]. Jiang et al. (2013) also observed that the increase in oxygen availability in the upflow biofilter resulted in utilization by nitrifiers rather than heterotrophic biomass [19].

We suggest that the typical low biodegradability of the organic carbon at the secondary effluent, which was also reflected in the low removal rate of the COD, BOD, and $UVA_{254}$, mentioned earlier (Section 3.4, Figure A5a,b in the Appendix A) was a key parameter on the rates of heterotrophic activity [44].

The environmental conditions of a low degradability rate of the organic matter and the addition of oxygen via $H_2O_2$, could be a possible explanation for favoring $NH_4^+$ rather than organic material oxidation. Such conditions, may promote simultaneous nitrification/nitrogen-removal activity. The significant relative abundance of genes with the potential for $NO_2^-$ oxidation on one hand and $NO_2^-$ reduction on the other hand strongly suggest that competition for $NO_2^-$ is a key in the dynamics of the community structure and function.

To address the paradox of excess oxygen availability in the reactor, due to $H_2O_2$ addition and degradation, while the relative abundance of genes related to anoxic activity reflects conditions of limited oxygen, we hypothesize that this community structure reflects inner localization in the biofilm: Whereas AOB and NOB are localized at the outer layer of the biofilm, degrade the $H_2O_2$, and utilize the supplemental oxygen for nitrification, in parallel with the depletion of oxygen along the depth of the biofilm, $NO_2^-$ and $NO_3^-$ are reduced. It is also possible that the nitrifiers are located in the pockets inside the biofilm and the better mass transfer of $H_2O_2$ enables utilization in a deeper location. We also suggest that the bacterial community that was investigated in this study indicates maximized thermodynamic utilization of electron acceptors and donors, with significant rates of anoxic activity.

The limitation of the process performance under increasing loads of $NH_4^+$ might be due to a lack of oxygen, limited HRT, or reactor volume and should be further investigated. As mentioned in previous research [34] the biofilter system succeeded in decreasing the oxygen demand and the drop in the ORP in following SAT. We suggest that the design presented in this study could be beneficial for treating secondary effluents containing residual levels of $NH_4^+$ (few mg/L-N), which is required to be removed under discharge or reuse restrictions.

## 4. Discussion

We suggest a model to describe the biokinetics of tertiary wastewater biofiltration systems with the addition of $H_2O_2$. The model focuses on the soluble (not particulate) content, carbon content similar to the secondary effluent (low biodegradability, nitrogen content), and the significance of metabolism over assimilation (Appendix A.1 and Table A1). Generally, although some full reduction of $NO_3^-$ to $NH_4^+$ may occur simultaneously, it is neglected in the model.

The model suggest that the addition of $H_2O_2$ (Figure 5) promoted the rates of the following pathways: (i) nitritation and (ii) nitratation, which occurs at high rates, and (iii) biodegradable COD oxidation, which occurs at minor rates. In parallel, (v) $NO_2^-$ and (vi) $NO_3^-$ reduction via denitrification and (vii) are also stimulated.

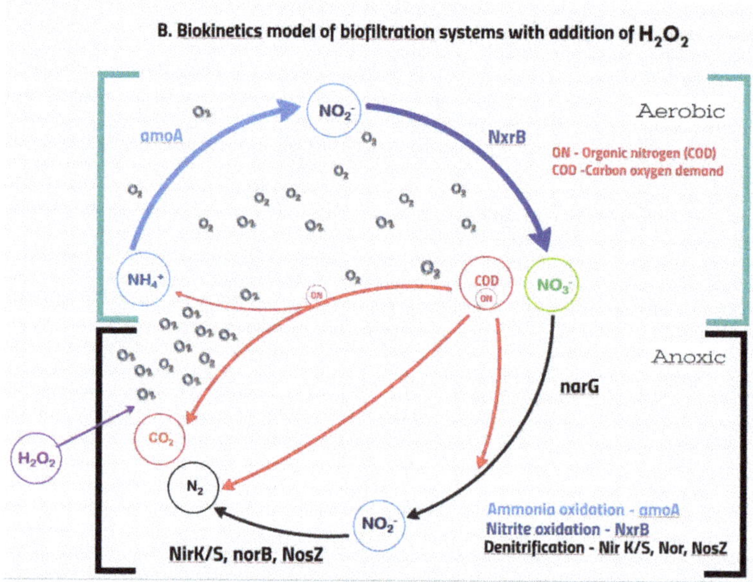


**Figure 5.** Suggested model for biokinetics of high-rate tertiary systems with the addition of $H_2O_2$. The arrows represent metabolic pathways with the relevant functional groups. Arrow thickness represents reaction rates relatively.

## 5. Conclusions

- The addition of $H_2O_2$ to stimulate aerobic activity within a bed filtration process, practiced under the low range of the high rate bed filtration (5–6 m/h), enabled a combined effect of particle filtration and nitrification.The $H_2O_2$ was fully degraded, limiting the nitrification rate.
- The trends in the concentration of $NH_4^+$, $NO_2^-$, and $NO_3^-$ with addition of $H_2O_2$ demonstrated significant nitirification activity at the bed filter.
- Metagenome analysis results were in line with the performance obtained and reflected significant potential for the simultanious nitrification denitrification activity of the attached biomass.
- The process presented herin may have a techno-economical potential, especially with the development of new and novel methods for the direct production of $H_2O_2$ [53,54] and the efficient application of the high-rate filtration.
- In cases of residual $NH_4^+$ concentration (<5 mg N-$NH_4^+$/L) in secondary WWTP effluents, the presented technology shows potential for either managing ammonia concentration in the effluent or for reducing oxygen demand in following processes, such as SAT or direct reuse, under local regulation.

**Author Contributions:** Conceptualization, D.A., H.M., M.J. and H.C.; methodology, U.H. and H.C., with design and building.; K.C. and J.S.-D., DNA sequencing, S.D. and M.E.; microbial sequencing analysis L.F. Conducting experiments writing—original draft preparation, writing—review, visualization and editing; All authors have read and agreed to the published version of the manuscript.

**Funding:** This research was funded by the BMBF-MOST (WT 1004) program and is a joint cooperation with TU Berlin, Mekorot, CU New York and TAU.

**Acknowledgments:** We thank Mekorot and Shafdan engineers Tomer Kraitzer and Avi Aharoni, for their contribution for successful R&D cooperation. Thanks to field engineer Eran Levi for dedicated and accurate work. We would also like to thank V. Cohen-Yaniv for her exact and clean work with the DNA extraction, Elk and his team from Cincinnati Children's Hospital for the DNA sequencing, S. Dagan for his work in the metagenome analysis.

**Conflicts of Interest:** The authors declare no conflict of interest.

## Appendix A

### Appendix A.1 Theoretical Calculation of and Oxygen Nitrogen Mass Balance

Nitrogen mass balance was calculated by making the following assumptions: (1) ammonia was the only nitrogen source for assimilation for biomass growth, (2) full aerobic utilization of oxygen originated from $H_2O_2$, (3) full nitrification to nitrate as the only electron acceptor for denitrification and (4) all nitrogen loss originated in anoxic denitrification activity. Two values for yields of 0.6 mg COD biomass/mg substrate COD (aerobic) as the upper limit and 0.4 mg COD biomass/mg substrate COD (anoxic) [43] as the lowest limit were used in the calculations as heterotroph activity was probably both aerobic and anoxic.

$$\text{N loss} = \text{BF}_{\text{Inlet}}\,[\text{mg/L TKN-N} + \text{mg/L } NO_2^{-}\text{N} + \text{mg/L } NO_3^{-}\text{N}] - \text{AF}_{\text{Outlet}}\,[\text{mg/L TKN-N} + \text{mg/L } NO_2^{-}\text{N} + \text{mg/L } NO_3^{-}\text{N}]. \quad \text{(A1)}$$

$$\text{OHO Nitrogen assimilated mg/L} = \Delta\text{COD} \times Y_{\text{OHO}} \times 14\ (\text{N gr/mole})/113\ (\text{Biomass gr/mole}) \quad \text{(A2)}$$

$$N_{\text{loss}}\ \text{denitrification} = \sum \Delta N_{\text{loss}} - \text{Nitrogen assimilated mg/L} \quad \text{(A3)}$$

$$\text{Denitrification credit} = N_{\text{loss}}\ \text{denitrification} \times 2.8\ \text{mg COD/mg } NO_3\text{-N [43]} = \Delta\text{COD}_{\text{denitrification}} \quad \text{(A4)}$$

The data presents the delta between the inlet and outlet: $\Delta NH_4^+$, $\Delta NO_2^-$, $\Delta NO_3^-$, $\Delta$COD, $\Delta\sum$N including Organic nitrogen (TKN).

Oxygen mass balance was calculated by making the following assumptions:

(1) All oxygen transformations were biological.

(2) All the degradation of $H_2O_2$ was biologically with oxygen and water as the only by products (based on previous lab calibrations).

(3) Excess oxygen, over the dissolved oxygen, in the inlet water was originated from $H_2O_2$ degradation.

$$\text{Degradation of 27 mg/L } H_2O_2 = 12.7\ \text{mg } O_2\ (1/0.47) \quad \text{(A5)}$$

$$O_2\ \text{demand for full nitrification} = 4.57\ \text{mg } O_2/\text{mg-N} \quad \text{(A6)}$$

$$O_2\ \text{mass balance}\ \Delta O_2 = (DO_{\text{inlet}} + 27\ \text{mg/L } H_2O_2 \times 0.47) - DO_{\text{outlet}} \quad \text{(A7)}$$

$$\text{Potential oxygen consumed for nitrification} = \Delta O_2 - O_2\ \textbf{demand for nitrification}\ (\Delta NH_4 \times 4.57) \quad \text{(A8)}$$

The theoretical demand of $H_2O_2$/mg-N $NH_4$ for full nitrification was calculated as follows:

$$C_{H2O2} \times 0.47\ (\text{mg } O_2/\text{mg } H_2O_2) \times 4.57\ (\text{mg } O_2/\text{mg-N}) = 9.72\ \text{mg of } H_2O_2/\text{mg-N } NH_4 \quad \text{(A9)}$$

A significant difference was obtained between the calculated concentrations of nitrogen assimilated with or without $H_2O_2$, in both estimates (0.8–1.15 mg/L-N vs 0.17–0.25 mg/L-N). Non-assimilative N loss (denitrification) was also significant with $H_2O_2$ (3–4 mg/L-N) while the calculated values of the control were within an error of measurement. This indicated significant enhancement of microbial growth and anoxic activity by $H_2O_2$ addition.

Considering the difference between the ammonia removed (theoretically oxidized to the oxygen balance it can be clearly seen that more $NH_4^+$ was consumed than oxygen was available along the increase in $NH_4^+$ concentration (negative values of mass balance).

**Table A1.** Nitrogen mass balance of the biofilter system with and without $H_2O_2$.

| | # Campaign | N Loss (mg-N/L) | OHO Assimilation (mg-N/L) | N Loss for Denitrification (mg-N/L) | Denitrification Credit (COD mg/L Utilized for Denitrification) |
|---|---|---|---|---|---|
| **OHO Yield = 0.6** | **0** | 1.52 | 1.04 | 0.48 | 1.3 |
| | *6* | 1.32 | 0.82 | 0.50 | 1.4 |
| | *15* | 1.43 | 1.04 | 0.39 | 1.1 |
| | **17** | 1.97 | 1.26 | 0.71 | 2.0 |
| | *20* | 3.71 | 1.64 | 2.07 | 5.8 |
| | *21* | 3.47 | 1.12 | 2.36 | 6.6 |
| | Average $H_2O_2$ | 2.24 | 1.15 | 1.09 | 3.0 |
| | STD dev. $H_2O_2$ | 0.98 | 0.25 | 0.81 | 2.27 |
| | **28** | 0.32 | 0.30 | 0.02 | 0.1 |
| | **29** | 0.25 | 0.15 | 0.10 | 0.3 |
| | **30** | 0.18 | 0.30 | 0.00 | 0.0 |
| | Average Control | 0.25 | 0.25 | 0.04 | 0.1 |
| | STD dev Control | 0.06 | 0.07 | 0.04 | 0.12 |
| | **# Campaign** | **N Loss (mg-N/L)** | **OHO Assimilation (mg-N/L)** | **N Loss for Deni (mg-N/L)** | **Denitrification Credit (COD mg/L Utilized for Deni)** |
| **OHO Yield = 0.4** | **0** | 1.52 | 0.69 | 0.83 | 2.3 |
| | *6* | 1.32 | 0.74 | 0.58 | 1.6 |
| | *15* | 1.43 | 0.69 | 0.74 | 2.1 |
| | **17** | 1.97 | 0.84 | 1.13 | 3.2 |
| | *20* | 3.71 | 1.09 | 2.62 | 7.3 |
| | *21* | 3.47 | 0.74 | 2.73 | 7.6 |
| | Average $H_2O_2$ | 2.24 | 0.80 | 1.44 | 4.0 |
| | STD $H_2O_2$ | 0.98 | 0.14 | 0.89 | 2.50 |
| | **28** | 0.32 | 0.20 | 0.12 | 0.3 |
| | **29** | 0.25 | 0.10 | 0.15 | 0.4 |
| | **30** | 0.18 | 0.20 | 0.00 | 0.0 |
| | Average Control | 0.25 | 0.17 | 0.09 | 0.3 |
| | STD dev. Control | 0.06 | 0.05 | 0.07 | 0.18 |

Notes: denitrification refers to denitrification.

**Table A2.** Oxygen mass Balance of the biofilter system with and without $H_2O_2$.

| # Campaign | $O_2$ Inlet (mg-/L) | $O_2$ Outlet (mg-/L) | $O_2$ from $H_2O_2$ | $O_2$ Balance (Consumed) | Delta $NH_4$ | Potential Oxygen Consumed for Nitrification |
|---|---|---|---|---|---|---|
| 1 | 3.79 | 5.1 | 12.7 | 11.39 | 0.98 | 6.92 |
| 2 | | | 12.7 | 12.70 | 1.35 | 6.53 |
| 3 | 5.53 | 4.83 | 12.7 | 13.40 | 1.54 | 6.37 |
| 4 | 5 | 4.72 | 12.7 | 12.98 | 1.85 | 4.53 |
| 5 | 3.87 | 3.62 | 12.7 | 12.95 | 2.55 | 1.30 |
| 6 | 3.8 | 4.12 | 12.7 | 12.40 | 2.63 | 0.37 |
| 7 | 3.43 | 3.18 | 12.7 | 12.95 | 2.75 | 0.38 |
| 8 | 6.30 | 5.17 | 12.7 | 13.83 | 2.77 | 1.16 |
| 9 | 5.38 | 4.42 | 12.7 | 13.66 | 3.05 | −0.28 |
| 10 | | | 12.7 | 12.70 | 3.00 | −1.01 |
| 11 | 4.69 | 3.92 | 12.7 | 13.47 | 3.30 | −1.61 |
| 12 | 3.92 | 3.75 | 12.7 | 12.87 | 3.55 | −3.35 |
| 13 | 5.83 | 5.2 | 12.7 | 13.32 | 3.56 | −2.93 |
| 14 | 4.1 | 4.49 | 12.7 | 12.26 | 4.08 | −6.39 |
| 15 | 6.15 | 4.27 | 12.7 | 14.58 | 4.35 | −5.30 |
| 16 | 5.13 | 3.53 | 12.7 | 14.30 | 3.60 | −2.15 |
| 17 | 4.50 | 4.41 | 12.7 | 12.79 | 4.32 | −6.94 |
| 18 | 3.83 | 4.16 | 12.7 | 12.37 | 4.46 | −8.00 |
| 19 | 5.67 | 4.10 | 12.7 | 14.27 | 3.90 | −3.55 |
| 20 | 5.48 | 4.28 | 12.7 | 13.90 | 4.12 | −4.94 |

### *Appendix A.2 The Backwash Cycle*

Periodic backwashing included combining air/water followed by water backwashing (Figure 1). The backwash cycle consisted of a short backwash to release filter-bed clogging by the air–water backwash, and another 10 min of water backwash and downflow wash to the sewer before restarting the filter. Backwashing volume was ~1.3% of the total volume filtered. The bottom plate nozzles were removed to improve the mixing efficiency and a side stream was also recirculated to the bottom part thus forming a sludge blanket which improved flocculation (verified by PSD analysis). The head loss was limited by online control to 0.5 bar/cycle and in the case of higher head loss, a second automatic backwash was performed. The quality of filtration was observed by the stable performance with low outlet turbidity (0.4–0.8 NTU).

### *Appendix A.3 Water Parameters Methodology*

The water quality of the samples was conducted in the SHAFDAN lab using selective electrode 4500-$NH_3$ for ammonium, 4500-$NO_2$-B colorimetric method for nitrite, 4110B ion chromatography for nitrate with chemical suppression of eluent conductivity, 5910B ultraviolet absorption method for UV absorbance at 254 nm (UVA), 5310B high-temperature combustion method for dissolved organic carbon (DOC), 4500-O C azide modification for DO, 5220B open reflux method for chemical oxygen demand (COD) and NOVA 60 kit for $H_2O_2$.

*Appendix A.4 $H_2O_2$ Decomposition*

Generally, the reaction occurs in two distinct stages in the catalytic reaction pathway. First, oxidation of the enzyme by one molecule of $H_2O_2$, while the oxygen–oxygen bond in peroxide is cleaved. At the second stage, another molecule of $H_2O_2$ reacts with the bond oxygen ion and forms molecular oxygen and water [28].

*Appendix A.5 Secondary Effluent Particle Analysis*

Effluent particles and aggregates were analyzed by Micro Flow Imaging. In brief, a fluid sample is drawn through a flow cell, illuminated with a light-emitting diode at 470 nm, and the magnified image is captured by digital camera. This image is then automatically analyzed to determine the particles' equivalent circular diameter (ECD) which represents the diameter of a sphere that occupies the same two-dimensional surface area as the particle. Analysis was conducted on particles of between ~2 and 400 μm.

*Appendix A.6 Particles Distribution*

A typical PSD analysis of the secondary effluent after 500 μm, post flocculation and after the biofilter is presented in Figure A1.

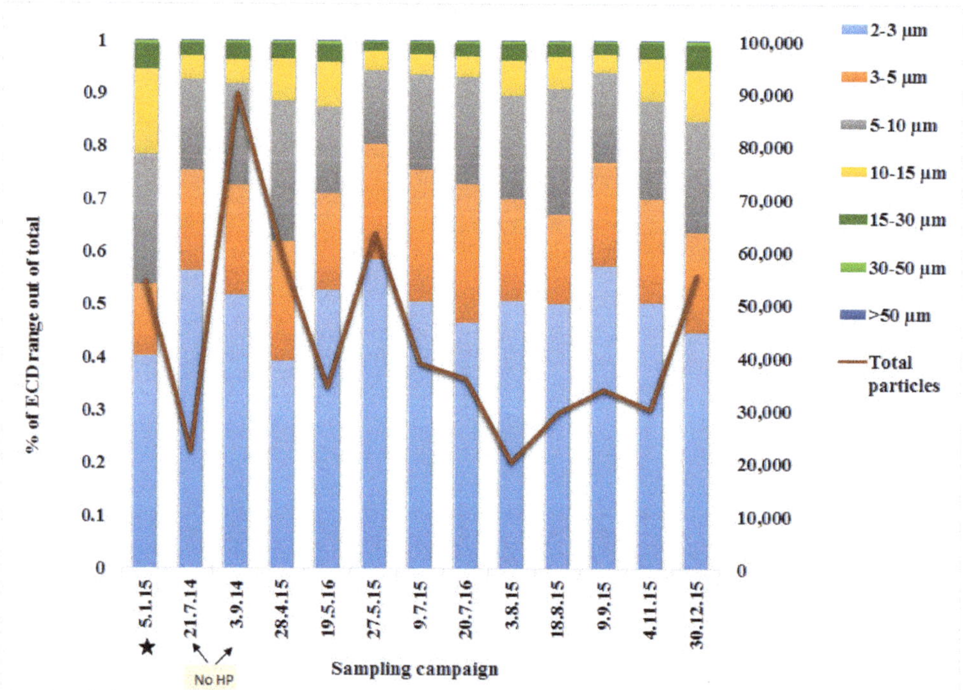


**Figure A1.** Particle distribution analysis over sampling campaigns. The Particle size is represented as an equivalent circular diameter (ECD). Total particle concentration (#/mL) is plotted as a red line. The results presented at Figure 2 refer to the campaign marked with a star. No HP is related to control runs without the addition of $H_2O_2$.

*Appendix A.7 Media Sampling and Metagenome Analysis*

The metagenome analysis was conducted to investigate the functional potential of the microbial media community. At the middle of the operation cycle (6 h) the biofilter was emptied and a sample of 50 g anthracite media was taken from a depth of 40 cm from the top of the media, thoroughly mixed for homogenization and immediately stored in dry ice. DNA from the media was extracted using Exgene Soil DNA Mini Kit. Whole-genome (shotgun) libraries were prepared using the TruSeq DNA PCR-Free HT Library Preparation Kit (Illumina, San Diego, CA, USA) and sequenced using an Illumina MiSeq benchtop sequencer with a target fragment length of 250 bp.

The final outputs of the metagenomics sequencing were modified using Cutadapt [55] to remove adapters and low-quality reads, and contigs were created using IDBA-UD [56]. A total of 8.9 M reads, was produced. The construction of a non-redundant gene catalog and the quantification of reference gene abundance (Gene calling) on the assembled scaftigs was performed using Prodigal [57]. Predicted genes were clustered using UCLUST [58]. High-quality reads were mapped to the reference gene catalog using BWA [59]. Mapped reads were used to form an abundance vector of the number of reads mapped to each gene. This was normalized based on data set size and gene length. Annotation was done using USEARCH [58] against the NCBI database (best hit with Eo1e-5, Bitscore >60 and sequence similarity >30%). Genus annotation was done using BWA aligner against the entire NCBI genome database (~80,000 genomes) and using PathoScope 2.0 [60] to quantify proportions of contigs from individual strains.

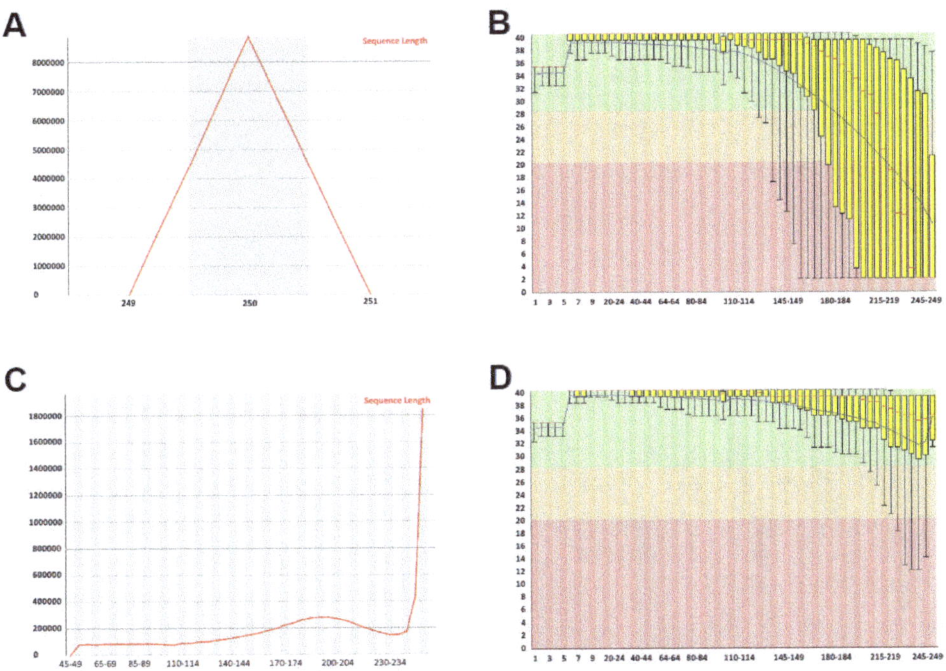


**Figure A2.** Metagenomics quality and length distribution: (**A**,**B**) before applying Cutadapt (Martin 2017): length distribution (**A**) and box plot distribution of quality score (Phred) (**B**). (**C**,**D**)—length and Phred distribution after applying Cutadapt.

*Appendix A.8 Functional Molecular Analysis of the Biomass*

The diversity of species within the sample indicating low diversity of the community in the sample. The number of reads that were aligned to nitrogen-related species were 43% of the total aligned reads, reflecting significant functional potential for the nitrogen cycle in the sampled bacterial community. Gene catalog was extracted from NCBI's nucleotide database. Mapped reads were subsequently filtered by removing those with a poor alignment (−q30). The remaining mapped reads were used to form an abundance matrix of the number of reads mapped to each gene in every sample. The abundance matrix was normalized based on data set.

Genes aligning to the reads were sorted according to their respective identified organisms which varied between 6%–19% among all the reads related to each gene (Figure A3).

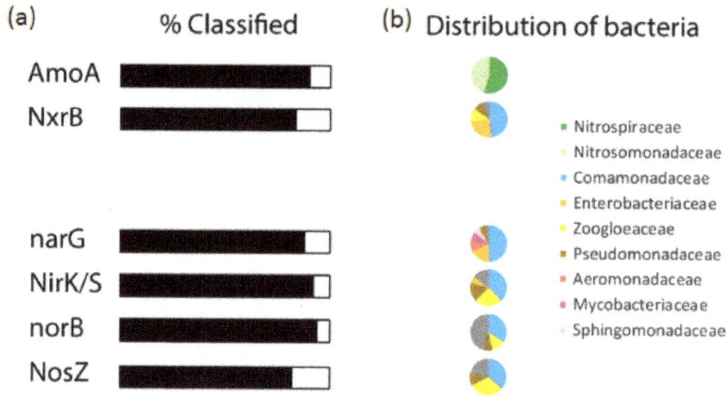


**Figure A3.** Percentage of classified (**white**) and unclassified (**black**) reads out of all the reads of each gene (**a**) and the composition of the identified genus of each gene (**b**).

*Appendix A.9 Biofilter Performance—Particle Analysis*

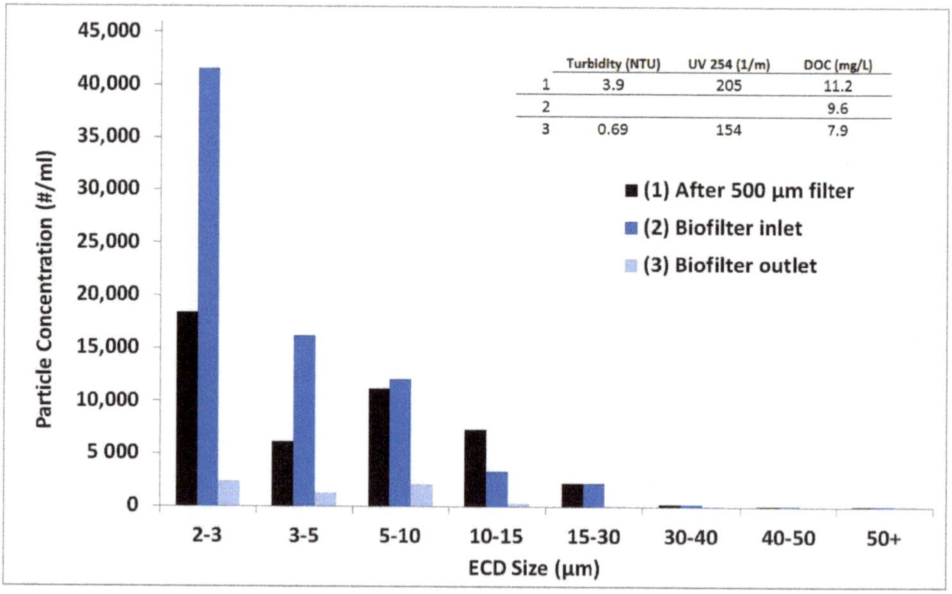


**Figure A4.** Particle-size distribution (PSD) analysis of a representative sampling campaign with $H_2O_2$. A comparison between sampling points after 500µm filtration (1) at the inlet of the biofilter after coagulation and flocculation (2) and at the outlet of the biofilter (3).

Flocculation before biofiltration increased particle counts, due to flocculation of the macromolecules and colloid particles (<2 μm) which were not analyzed. As expected, particle concentration decreased dramatically after biofiltration for all ECDs.

*Appendix A.10 Organic Carbon Performance*

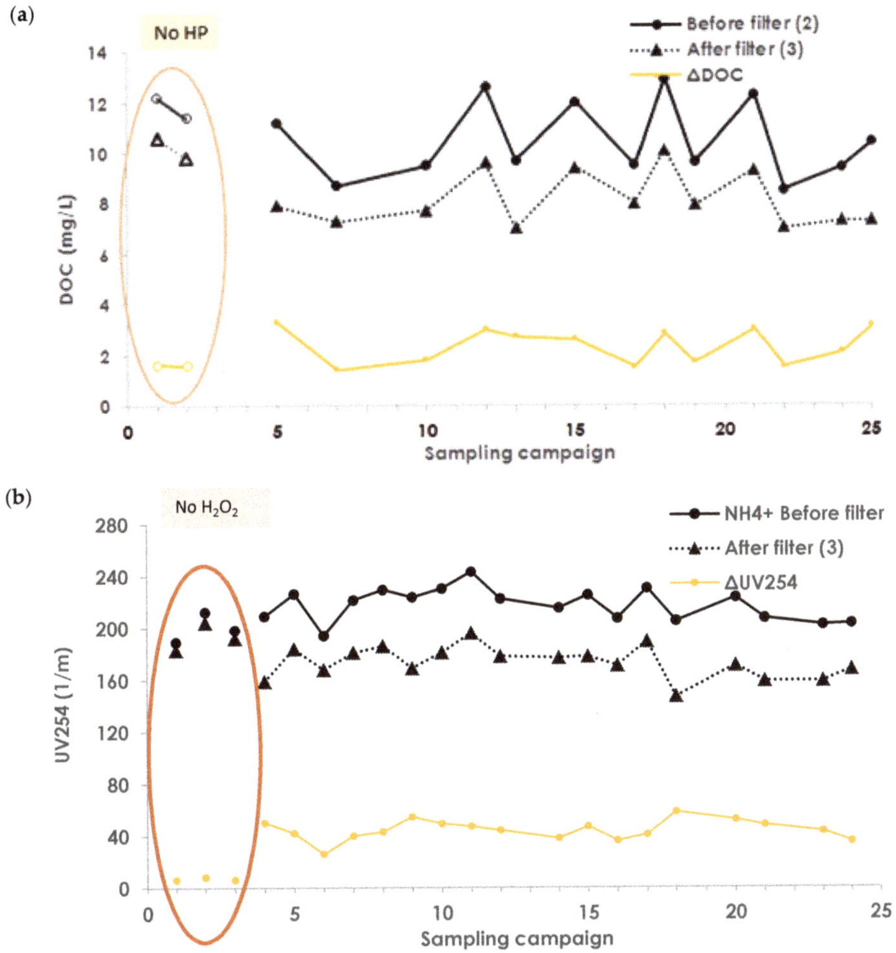


**Figure A5.** Concentrations and the removal of DOC (**a**) and UVA ($UV_{254}$) (**b**) over sampling campaigns before the biofilter (BF) and after the biofilter (AF), which was also sorted by feed values of ammonium during experiments with $H_2O_2$ addition. Reference campaigns of no $H_2O_2$ are circled in orange.

*Appendix A.11 Nitrite Removal and Concentration*

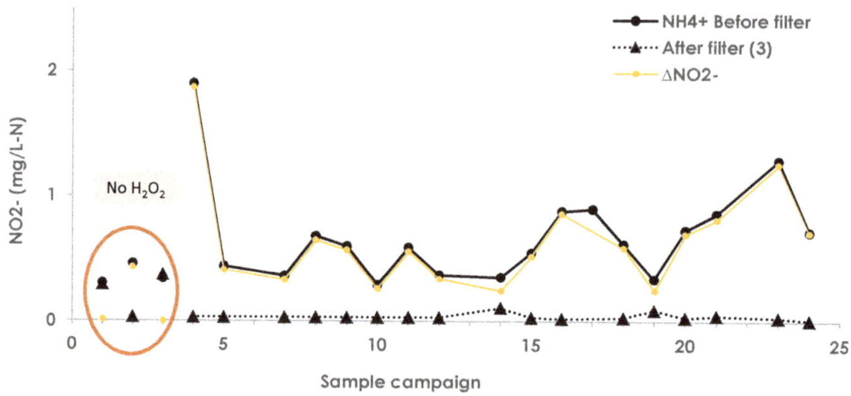


**Figure A6.** Concentrations and removal of nitrite (mg/L-N) over sampling campaigns, before the biofilter (BF) and after the biofilter (AF) with $H_2O_2$ addition, also sorted by feed values of ammonium. No $H_2O_2$ campaigns are circled in orange.

## References

1. Bixio, D.; De Hayder, B.; Cikurel, H.; Muston, M.; Miska, V.; Joksimovic, D.; Schäfer, A.I.; Ravazzini, A.; Aharoni, A.; Savic, D.; et al. Municipal wastewater reclamation: Where do we stand? An overview of treatment technology and management practice. *Water Sci. Technol. Water Supply* **2005**, *5*, 77–86. [CrossRef]
2. Siegel, S.M. *Let There Be Water: Israel's Solution for a Water-Starved World*; Thomas Dunne Books/St. Martin's Press: New York, NY, USA, 2015; ISBN 978-1-250-07395-2.
3. *Emerging Technologies for Wastewater Treatment and In-Plant Wet Weather Management*; U.S Environmental Protection Agency (EPA): Washington DC, USA, 2013; Volume 832.
4. Tchobanoglous, F.B.G. *Wastewater Engineering and Reuse*, 4th ed.; McGraw-Hill: New York, NY, USA, 2003; ISBN 978-1-259-01079-8.
5. Van Den Akker, B.; Holmes, M.; Cromar, N.; Fallowfield, H. Application of high rate nitrifying trickling filters for potable water treatment. *Water Res.* **2008**, *42*, 4514–4524. [CrossRef]
6. Miska-Markusch, V. *Effluent Filtration for more than Particle Removal*; Leibniz Universität Hannover: Hannover, Germany, 2009.
7. Müller, J.; Drewes, J.E.; Hübner, U. Sequential biofiltration—A novel approach for enhanced biological removal of trace organic chemicals from wastewater treatment plant effluent. *Water Res.* **2017**, *127*, 127–138. [CrossRef]
8. Alidina, M.; Li, D.; Drewes, J.E. Investigating the role for adaptation of the microbial community to transform trace organic chemicals during managed aquifer recharge. *Water Res.* **2014**, *56*, 172–180. [CrossRef]
9. Palomo, A.; Fowler, S.J.; Gülay, A.; Rasmussen, S.; Sicheritz-Ponten, T.; Smets, B.F. Metagenomic analysis of rapid gravity sand filter microbial communities suggests novel physiology of Nitrospira spp. *ISME J.* **2016**, *10*, 2569–2581. [CrossRef]
10. Pinto, A.J.; Marcus, D.N.; Ijaz, Z.; Bautista-de los Santos, Q.M.; Dick, G.J.; Raskin, L. Metagenomic Evidence for the Presence of Comammox Nitrospira-Like Bacteria in a Drinking Water System. *mSphere* **2015**, *1*, e00054-15. [CrossRef]
11. Salerno, C.; Berardi, G.; Laera, G.; Pollice, A. Functional Response of MBR Microbial Consortia to Substrate Stress as Revealed by Metaproteomics. *Microb. Ecol.* **2019**, *78*, 873–884. [CrossRef]
12. Salerno, C.; Benndorf, D.; Kluge, S.; Palese, L.L. Metaproteomics Applied to Activated Sludge for Industrial Wastewater Treatment Revealed a Dominant Methylotrophic Metabolism of *Hyphomicrobium zavarzinii*. *Microb. Ecol.* **2016**, *72*, 9–13. [CrossRef]

13. Mendoza-Espinoza, L.; Stephenson, T. A Review of Biological Aerated Filters (BAFs) for Wastewater Treatment. *Environ. Eng. Sci.* **1999**, *16*, 201–216. [CrossRef]
14. Rassamee, V.; Sattayatewa, C.; Pagilla, K.; Chandran, K. Effect of oxic and anoxic conditions on nitrous oxide emissions from nitrification and denitrification processes. *Biotechnol. Bioeng.* **2011**, *108*, 2036–2045. [CrossRef]
15. Yoo, H.; Ahn, K.; Lee, H.J.; Lee, K.; Kwak, Y.; Song, K. Nitrogen removal from synthetic wastewater by simultaneous nitrification and denitrification (SND) via nitrite in an intermittently-aerated reactor. *Water Res.* **1999**, *33*, 145–154. [CrossRef]
16. Bassin, J.P.; Abbas, B.; Vilela, C.L.S.; Kleerebezem, R.; Muyzer, G.; Rosado, A.S.; van Loosdrecht, M.C.M.; Dezotti, M. Tracking the dynamics of heterotrophs and nitrifiers in moving-bed biofilm reactors operated at different COD/N ratios. *Bioresour. Technol.* **2015**, *192*, 131–141. [CrossRef]
17. de Vet, W.W.J.M.; Kleerebezem, R.; van der Wielen, P.W.J.J.; Rietveld, L.C.; van Loosdrecht, M.C.M. Assessment of nitrification in groundwater filters for drinking water production by qPCR and activity measurement. *Water Res.* **2011**, *45*, 4008–4018. [CrossRef]
18. Yu, X.; Qi, Z.; Zhang, X.; Yu, P.; Liu, B.; Zhang, L.; Fu, L. Nitrogen loss and oxygen paradox in full-scale biofiltration for drinking water treatment. *Water Res.* **2007**, *41*, 1455–1464. [CrossRef]
19. Jiang, B.; Hu, W.; Pei, H.; Chen, P.; Liu, Q. The influence of aeration on nitrification and the nitrifier distribution in an upflow biological aerated filter for tertiary treatment of municipal sewage. *Desalin. Water Treat.* **2010**, *24*, 308–320. [CrossRef]
20. Albuquerque, A.; Makinia, J.; Pagilla, K. Impact of aeration conditions on the removal of low concentrations of nitrogen in a tertiary partially aerated biological filter. *Ecol. Eng.* **2012**, *44*, 44–52. [CrossRef]
21. Matsumoto, S.; Terada, A.; Tsuneda, S. Modeling of membrane-aerated biofilm: Effects of C/N ratio, biofilm thickness and surface loading of oxygen on feasibility of simultaneous nitrification and denitrification. *Biochem. Eng. J.* **2007**, *37*, 98–107. [CrossRef]
22. Brindle, K.; Stephenson, T.; Semmens, M.J. Nitrification and oxygen utilisation in a membrane aeration bioreactor. *J. Memb. Sci.* **1998**, *144*, 197–209. [CrossRef]
23. Pellicer-Nàcher, C.; Smets, B.F. Structure, composition, and strength of nitrifying membrane-aerated biofilms. *Water Res.* **2014**, *57*, 151–161. [CrossRef]
24. Fiorenza, S.; Ward, C.H. Microbial adaptation to hydrogen peroxide and biodegradation of aromatic hydrocarbons. *J. Ind. Microbiol. Biotechnol.* **1997**, *18*, 140–151. [CrossRef]
25. Zappi, M.; White, K.; Hwang, H.-M.; Bajpai, R.; Qasim, M. The Fate of Hydrogen Peroxide as an Oxygen Source for Bioremediation Activities within Saturated Aquifer Systems. *J. Air Waste Manag. Assoc.* **2000**, *50*, 1818–1830. [CrossRef]
26. Chance, B.; Herbert, D. The enzyme-substrate compounds of bacterial catalase and peroxides. *Biochem. J.* **1950**, *46*, 402–414. [CrossRef]
27. Northrop, J.H. Experimental Conditions and General Results. Effect of Varying the Concentration of Catalase. *J. Gen. Physiology* **1924**, *5*, 373–387.
28. Sooch, B.S.; Kauldhar, B.S.; Puri, M. Recent insights into microbial catalases: Isolation, production and purification. *Biotechnol. Adv.* **2014**, *32*, 1429–1447. [CrossRef]
29. Flores, M.J.; Brandi, R.J.; Cassano, A.E.; Labas, M.D. Chemical disinfection with $H_2O_2$—The proposal of a reaction kinetic model. *Chem. Eng. J.* **2012**, *198–199*, 388–396. [CrossRef]
30. Lakretz, A.; Ron, E.Z.; Mamane, H. Biofilm control in water by a UV-based advanced oxidation process. *Biofouling* **2011**, *27*, 295–307. [CrossRef]
31. Tusseau-Vuillemin, M.H.; Lagarde, F.; Chauviere, C.; Heduit, A. Hydrogen peroxide ($H_2O_2$) as a source of dissolved oxygen in COD-degradation respirometric experiments. *Water Res.* **2002**, *36*, 793–798. [CrossRef]
32. Lakretz, A.; Ron, E.Z.; Harif, T.; Mamane, H. Biofilm control in water by advanced oxidation process (AOP) pre-treatment: Effect of natural organic matter (NOM). *Water Sci. Technol.* **2011**, *64*, 1876–1884. [CrossRef]
33. Yang, Y.; Cheng, D.; Li, Y.; Yu, L.; Gin, K.Y.H.; Chen, J.P.; Reinhard, M. Effects of monochloramine and hydrogen peroxide on the bacterial community shifts in biologically treated wastewater. *Chemosphere* **2017**, *189*, 399–406. [CrossRef]
34. Zucker, I.; Mamane, H.; Cikurel, H.; Jekel, M.; Hubner, U.; Avisar, D. A hybrid process of biofiltration of secondary effluent followed by ozonation and short soil aquifer treatment for water reuse. *Water Res.* **2015**, *84*, 315–322. [CrossRef]

35. Cikurel, H.; Rebhun, M.; Amirtharajah, A.; Adin, A. Wastewater effluent reuse by in-line flocculation filtration process. *Water Sci. Technol.* **1996**, *33*, 203–211. [CrossRef]
36. Mamane, H.; Kohn, C.; Adin, A. Characterizing Shape of Effluent Particles by Image Analysis. *Sep. Sci. Technol.* **2008**, *43*, 1737–1753. [CrossRef]
37. Li, D.; Sharp, J.O.; Saikaly, P.E.; Ali, S.; Alidina, M.; Alarawi, M.S.; Keller, S.; Hoppe-Jones, C.; Drewes, J.E. Dissolved organic carbon influences microbial community composition and diversity in managed aquifer recharge systems. *Appl. Environ. Microbiol.* **2012**, *78*, 6819–6828. [CrossRef] [PubMed]
38. Lauderdale, C.; Chadik, P.; Kirisits, M.J.; Brown, A.J. Engineered biofiltration: Enhanced biofilter performance through nutrient and peroxide addition. *J. Am. Water Work. Assoc.* **2012**, *104*, E298–E309. [CrossRef]
39. Lakretz, A.; Elifantz, H.; Kviatkovski, I.; Eshel, G.; Mamane, H. Automatic microfiber filtration (AMF) of surface water: Impact on water quality and biofouling evolution. *Water Res.* **2014**, *48*, 592–604. [CrossRef]
40. Li, H.; Durbin, R. Fast and accurate short read alignment with Burrows-Wheeler transform. *Bioinformatics* **2009**, *25*, 1754–1760. [CrossRef]
41. Shon, H.K.; Vigneswaran, S.; Snyder, S.A. Effluent Organic Matter (EfOM) in Wastewater: Constituents, Effects, and Treatment. *Crit. Rev. Environ. Sci. Technol.* **2006**, *36*, 327–374. [CrossRef]
42. Petala, M.; Tsiridis, V.; Samaras, P.; Zouboulis, A.; Sakellaropoulos, G.P. Wastewater reclamation by advanced treatment of secondary effluents. *Desalination* **2006**, *195*, 109–118. [CrossRef]
43. Grady, C.P.L.; Daigger, G.T.; Love, N.G.; Filipe, C.D.M. *Biological Wastewater Treatment*; IWA Publishing—Co-Publication: London, UK, 2011; ISBN 9781843393429.
44. Friedman, L.; Mamane, H.; Avisar, D.; Chandran, K. The role of influent organic carbon-to-nitrogen (COD/N) ratio in removal rates and shaping microbial ecology in soil aquifer treatment (SAT). *Water Res.* **2018**, *146*, 197–205. [CrossRef]
45. Cua, L.S.; Stein, L.Y. Effects of nitrite on ammonia-oxidizing activity and gene regulation in three ammonia-oxidizing bacteria. *FEMS Microbiol. Lett.* **2011**, *319*, 169–175. [CrossRef]
46. Chen, J.; Strous, M. Denitrification and aerobic respiration, hybrid electron transport chains and co-evolution. *Biochim. Biophys. Acta Bioenerg.* **2013**, *1827*, 136–144. [CrossRef] [PubMed]
47. Daims, H.; Lücker, S.; Wagner, M. A New Perspective on Microbes Formerly Known as Nitrite-Oxidizing Bacteria. *Trends Microbiol.* **2016**, *24*, 699–712. [CrossRef] [PubMed]
48. Kits, K.D.; Sedlacek, C.J.; Lebedeva, E.V.; Han, P.; Bulaev, A.; Pjevac, P.; Daebeler, A.; Romano, S.; Albertsen, M.; Stein, L.Y.; et al. Kinetic analysis of a complete nitrifier reveals an oligotrophic lifestyle. *Nature* **2017**, *549*, 269–272. [CrossRef] [PubMed]
49. Jetten, M.S.M. New pathways for ammonia conversion in soil and aquatic systems. *Plant Soil* **2001**, *230*, 9–19. [CrossRef]
50. Stein, L.Y. Heterotrophic nitrification and nitrifier denitrification. In *Nitrification*; American Society of Microbiology: Washington, DC, USA, 2011.
51. Ge, S.; Wang, S.; Yang, X.; Qiu, S.; Li, B.; Peng, Y. Detection of nitrifiers and evaluation of partial nitrification for wastewater treatment: A review. *Chemosphere* **2015**, *140*, 85–98. [CrossRef] [PubMed]
52. Park, M.R.; Park, H.; Chandran, K. Molecular and kinetic characterization of planktonic *Nitrospira* spp. selectively enriched from activated sludge. *Environ. Sci. Technol.* **2017**, *51*, 2720–2728. [CrossRef] [PubMed]
53. Siahrostami, S.; Verdaguer-Casadevall, A.; Karamad, M.; Deiana, D.; Malacrida, P.; Wickman, B.; Escudero-Escribano, M.; Paoli, E.A.; Frydendal, R.; Hansen, T.W.; et al. Enabling direct $H_2O_2$ production through rational electrocatalyst design. *Nat. Mater.* **2013**, *12*, 1137–1143. [CrossRef]
54. Yang, S.; Verdaguer-casadevall, A.; Arnarson, L.; Silvioli, L.; Viktor, C.; Frydendal, R.; Rossmeisl, J.; Chorkendor, I.; Stephens, I.E.L. Toward the Decentralized Electrochemical Production of $H_2O_2$: A Focus on the Catalysis. *ACS Catal.* **2018**, *8*, 4064–4081. [CrossRef]
55. Martin, M. Cutadapt removes adapter sequences from high-throughput sequencing reads. *EMBnet. J.* **2011**, *17*, 10–12. [CrossRef]
56. Peng, Y.; Leung, H.C.M.; Yiu, S.M.; Chin, F.Y.L. IDBA-UD: A de novo assembler for single-cell and metagenomic sequencing data with highly uneven depth. *Bioinformatics* **2012**, *28*, 1420–1428. [CrossRef]
57. Hyatt, D.; Chen, G.-L.; LoCascio, P.F.; Land, M.L.; Larimer, F.W.; Hauser, L.J. Prodigal: Prokaryotic gene recognition and translation initiation site identification. *BMC Bioinform.* **2010**, *11*, 119. [CrossRef] [PubMed]
58. Edgar, R.C. Search and clustering orders of magnitude faster than BLAST. *Bioinformatics* **2010**, *26*, 2460–2461. [CrossRef]

59. Li, H.; Durbin, R. Fast and accurate long-read alignment with Burrows-Wheeler transform. *Bioinformatics* **2010**, *26*, 589–595. [CrossRef] [PubMed]
60. Hong, C.; Manimaran, S.; Shen, Y.; Perez-Rogers, J.F.; Byrd, A.L.; Castro-Nallar, E.; Crandall, K.A.; Johnson, W.E. PathoScope 2.0: A complete computational framework for strain identification in environmental or clinical sequencing samples. *Microbiome* **2014**, *2*, 33. [CrossRef] [PubMed]

clean technologies

MDPI

*Review*

# Removal of Odors (Mainly $H_2S$ and $NH_3$) Using Biological Treatment Methods

**Krzysztof Barbusiński [1], Anita Parzentna-Gabor [1,2,*] and Damian Kasperczyk [2]**

1 Department of Water and Wastewater Engineering, Silesian University of Technology, Konarskiego 18, 44-100 Gliwice, Poland; krzysztof.barbusinski@polsl.pl
2 Ekoinwentyka Ltd., Szyb Walenty 26, 41-700 Ruda Śląska, Poland; biuro@ekoinwentyka.pl
* Correspondence: sekretariat@ekoinwentyka.pl

**Abstract:** This study reviews the available and most commonly used methods of gas deodorization. Comparing various methods of odor removal, undoubtedly biological methods of pollution degradation have an advantage over others—chemical and physical. This advantage is manifestedmainly in ecological and economic terms. The possibility of using biological methods to remove $H_2S$ and $NH_3$, as the most common emitted by the municipal sector companies, was analyzed in terms of their removal efficiency. The method of bio-purification of air in biotrickling filters is more advantageous than the others, due to the high effectiveness of VOCs and odors degradation, lack of secondary pollutants, and economic aspects—it is a method competitive to the commonly used air purification method in biofilters.

**Keywords:** biodegradation; odors; $H_2S$; $NH_3$

**Citation:** Barbusiński, K.; Parzentna-Gabor, A.; Kasperczyk, D. Removal of Odors (Mainly $H_2S$ and $NH_3$) Using Biological Treatment Methods. *Clean Technol.* **2021**, *3*, 138–155. https://doi.org/10.3390/cleantechnol3010009

Received: 19 November 2020
Accepted: 18 January 2021
Published: 17 February 2021



## 1. Introduction

Fragrances, also known as odors, and volatile organic compounds (VOCs) are compounds that play an important role in environmental pollution. They are gaseous pollutants and chemical vapors, characterized by high vapor pressure and low solubility in water. The presence of odors is a problem mainly in urbanized areas, whose source is the combustion of hydrocarbon fuels, chemical and petrochemical industries, mining, municipal, food processing, and agricultural waste [1]. The economic with the highest odor nuisance are primarily waste management plants and wastewater treatment plants, as well as food and agricultural processing plants. Odors are a mixture of volatile chemicals known as odorous gases that can be felt by humans in very low concentrations. They play an important role in the lives of all living organisms, especially humans, causing health problems. Odors are carcinogenic, mutagenic and toxic, and due to odor nuisance, they also affect mental health. These compounds cause irritation of the respiratory system, which results in runny nose, cough, shortness of breath, sore throat, and lacrimation. In many cases, due to people's individual hypersensitivity, psychosomatic symptoms, such as insomnia, reduced psychophysical and emotional performance, and panic attacks also occur [2,3]. The most exposed to the described health problems are people living in the immediate vicinity of the emission sources, as well as employees of odor emitting plants. The incoming new residential and industrial investments cause an increase in the density of development the urban agglomeration, and thus too close proximity to residential buildings with odor-emitting plants, especially wastewater treatment plants, which are very often located in city centers, in the vicinity of large housing estates [4].

## 2. Gases Emitted in the Municipal Sector

Facilities emitting the most persistent harmful gases include companies from the municipal sector, including waste water treatment plants, waste management plants or composting plants. The odor- produced in municipal wastewater treatment plants include

- sulfur compounds, i.e., hydrogen sulfide, thiols, sulfides, and alkyl disulfides,
- nitrogen compounds, i.e., ammonia, and aliphatic amines,
- organic compounds, including aldehydes, ketones, and fatty acids (phenol, cresol, butyric acid, acetic acid, and valeric acid)

Table 1 lists the most common and nuisance odor compounds found in municipal wastewater treatment plants.

**Table 1.** Concentrations of odor compounds above the surface of municipal wastewater [5] (Authors' own study according to [5], Polish Scientific Publishing House: 2012).

| Odor Compounds | Substance | Concentration [ppm] | Detection Threshold [ppm] |
|---|---|---|---|
| Compounds with Nitrogen | Ammonia | 0.019–5.2 | 5 |
| | Trimethylamine | 1.7 | 0.00044 |
| | Methylamine | 3.3 | 0.02 |
| | Pyridine | 0.013–0.82 | 0.084 |
| Compounds with Sulfur | Hydrogen sulfide | 0.001–0.78 | 0.008 |
| | Dimethyl sulfide | 0.0015–0.02 | 0.0023 |
| | Diethyl sulfide | 0.00025–0.0006 | 0.004 |
| | Diethyl disulfide | 0.000054 | 0.00043 |
| | Methyl mercaptan | 0.0001–0.55 | 0.001 |
| | Ethyl mercaptan | 0.000016–0.074 | 0.00076 |
| Volatile Organic Compounds | Phenol | 0.047–0.65 | 0.040 |
| | Cresol | 0.00047 | 0.0018 |
| | Butter acid | 0.00028–0.00056 | 0.004 |
| | Valeric acid | 0.0006 | 0.005 |

In most wastewater treatment plants, ammonia and hydrogen sulfide have the highest concentrations and amounts of odor compounds in the emitted gases [6,7]. Ammonia is a colorless, corrosive gas with a very pungent and unpleasant odor. Its corrosive and exothermic properties can damage the eyes, skin, and mucous membranes of the mouth and respiratory tract. The effect of ammonia on the human body and the accompanying disease symptoms are presented in Table 2.

**Table 2.** The effect of ammonia on the human body [8] (Authors' own study according to [8]. The National Academies Press: Washington: 2008).

| Concentration $NH_3$ [ppm] | Symptoms |
|---|---|
| <5–53 | Odor threshold |
| 30 | Slight irritation after 10 min. |
| 50 | Prolonged exposure may cause nausea, tearing of the eyes, Moderate irritation to the eyes, nose, throat and chest after 10 min to 2h |
| 80 | Moderate to highly irritation after 30 min to 2 h |
| 110 | Highly intense irritation after 30 min to 2hof exposure |
| 140 | Unbearable irritation after 30 min to 2h |
| 500 | Excessive lacrimation and irritation |
| 570 (21–30 years old)<br>1000 (60 years old)<br>1790 (86–90 years old) | Reflex glottis closure—a protective response to inhaling irritant vapors |

Another very nuisance odor compound, emitted by wastewater treatment plants, is hydrogen sulfide—a colorless, highly flammable and explosive gas, with the smell of rotten eggs, felt in very low concentrations. This gas has a toxic effect on all living organisms, including humans, causing a number of unpleasant life-threatening ailments. The effect of hydrogen sulfide on the human body along with disease symptoms is presented in Table 3.

**Table 3.** The effect of dihydrogen sulfide on the human body [9].

| Concentration $H_2S$ [ppm] | Symptoms |
|---|---|
| 0.00011–0.00033 | Typical background concentrations |
| 0.01–1.5 | Odor threshold (rotten egg) |
| 2–5 | Prolonged exposure may cause nausea, tearing of the eyes, headaches or loss of sleep |
| 20 | Possible fatigue, loss of appetite, headache, irritability, poor memory, dizziness |
| 50–100 | Slight conjunctivitis ("gas eye") and respiratory tract irritation after 1 h. May cause digestive upset and loss of appetite |
| 100 | Coughing, eye irritation, loss of smell after 2–15 min (olfactory fatigue). Altered breathing, drowsiness after 15–30 min. Throat irritation after 1 h. Gradual increase in severity of symptoms over several hours. Death may occur after 48 h |
| 100–150 | Loss of smell (olfactory fatigue or paralysis) |
| 200–300 | Marked conjunctivitis and respiratory tract irritation after 1 h. Pulmonary edema may occur from prolonged exposure |
| 500–700 | Staggering, collapse in 5 min. Serious damage to the eyes in 30 min. Death after 30–60min |
| 700–1000 | Rapid unconsciousness, "knockdown" or immediate collapse within 1 to 2 breaths, breathing stops, death within minutes |
| 1000–2000 | Nearly instant death |

The main factors influencing the concentration of odors in the air are the rate of emission and dispersion of gases, which are directly affected by atmospheric conditions (temperature, wind direction, and atmospheric pressure) and geomorphological conditions [10]. The most odor-nuisance areas in the wastewater treatment plant include pretreatment unit—raw wastewater tanks, coarse and fine screen, and sand traps—preliminary settling tanks and sludge management facilities—sludge tanks, and sludge dewatering halls [5,7,11]. Preliminary sedimentation tanks and sewage sludge tanks generate the largest amounts of odors due to the large area of these objects and the emission of gases from their entire surface [12]. Excessive sewage sludge generated in the wastewater treatment process is characterized by a high efficiency of rotting. For their disposal or further use, e.g., for land reclamation, it is necessary to stabilize the sludge and deprive it of pathogenic organisms. For this purpose, the excess sludge is directed to hermetic fermentation chambers, where the four-stage process of anaerobic fermentation takes place, with the participation of several groups of anaerobic bacteria. These bacteria break down complex organic substances (proteins, fats, and carbohydrates) into methane and carbon dioxide. In the fermentation process, the collected biogas can be used to produce heat and electricity [13,14] but before that, it must be purified [15]. Biogas resulting from fermentation consists mainly of methane and carbon dioxide as well as trace amounts of hydrogen, carbon monoxide, nitrogen, oxygen, and hydrogen sulfide (Figure 1), in order to be used for energy production in installations, it must meet the appropriate requirements adapted to such devices as engines and boilers, therefore it is necessary to clean them, also to increase their calorific value [16].

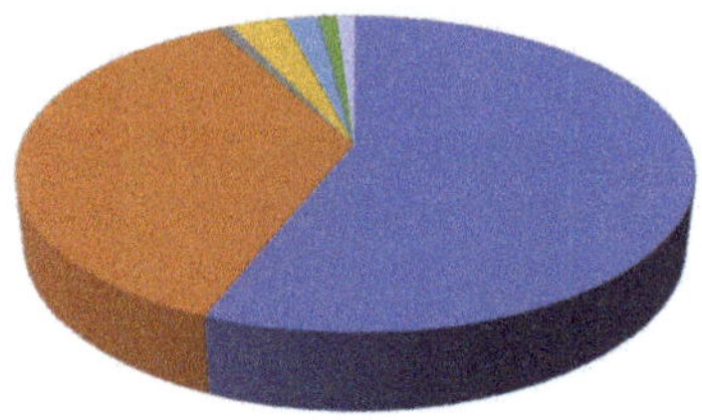


**Figure 1.** Composition of biogas [13,15].

Various techniques of biogas conditioning are used. The most frequently used are desulfurization by help of turf ore, membrane separation, pressure swing adsorption, physical absorption, chemical absorption, and biotechnological methods such as use of biofilters, bioscrubbers, biotrickling filters as well as activated sludge [15,17,18]. The selection of the appropriate technology for biogas purification depends on specific requirements, taking into account the type of installation and local conditions.

## 3. Commonly Used Odor Removal Technologies

Limiting odor emissions generated in wastewater treatment plant consists in preventing the emission of gases directly into the atmosphere, e.g., by hermitization the most odor-troublesome technological devices, and deodorization of exhaust gases. The most frequently used deodorization methods in municipal sector include absorption with the use of reactive oxidizing solutions, adsorption on activated carbon, combustion, and biological methods [7,19]. Recently, biological methods of odor removal have become more and more popular, which using natural reactions occurring in nature, are ecological, effective and inexpensive solutions [20].

### *3.1. Physicochemical Methods of Odor Removal*

The physicochemical methods for deodorizing gases emitted by wastewater treatment plants include absorption, adsorption, and combustion. Air purification by absorption method consists in transferring pollutants from the emitted gas to the liquid and enables the separation of the gas mixture into individual components [21]. In the case of odors from wastewater treatment plants, the absorption efficiency in water is very low, due to the low solubility of most odor pollutants.

In order to increase the efficiency of this process, solutions of oxidants are used as absorption liquids, e.g., ozone $O_3$, hydrogen peroxide $H_2O_2$, sodium chlorate (I) NaOCl, under the influence of which organic compounds are oxidized to carbon dioxide ($CO_2$), and hydrogen sulfide to elemental sulfur (S), mercaptans, and sulfides to sulfonic acids or sulfones. These reactions can be accelerated by adding appropriate catalysts (e.g., salts containing iron ions(II) $Fe^{2+}$) [22]. The use of reactive chemicals as absorption liquids requires the use of chemically resistant construction materials to minimize the risk of environmental contamination due to leakage of reagents. The absorption method is an effective solution for removing ammonia ($NH_3$) and hydrogen sulfide ($H_2S$), but it is much more difficult to remove volatile organic compounds, including volatile fatty acids, mercaptans. Moreover, this method generates noxious sewage that must be disposed of [5]. Chemical absorption in many cases is used as a pre-treatment method of emitted gases characterized by a high concentration of odors [7].

Adsorption is a process of inhibiting a pollutant by a solid—an adsorbent. Activated carbon and zeolite are most often used for deodorization, which are characterized by high adsorption capacity in relation to odor compounds [22]. Adsorbents used for deodorizing the emitted gases are in the form of powder (8–80 μm), granules (200 μm to 6 mm), compacts (0.8 to 5 mm in diameter and 5 to 20 mm long), pellets (30 to 60 mm in diameter), fibers or fabrics. Except activated carbon and zeolites, diatomaceous and volcanic earths, sawdust, silica, aluminum oxides, and peat are also used. In addition, clay minerals and polymeric synthetic resins are also used, but this group of adsorbents absorbs odorant molecules much worse. After complete saturation of the bed, its regeneration is carried out to remove adsorbed impurities, depending on their type, various methods are available: thermal, vacuum or chemical regeneration, storage, combustion, but in the case of deodorization, sorbent is usually not regenerated due to the risk of secondary odor emissions and small benefit [5].

In contrast, combustion can be generally divided into thermal and catalytic combustion. Thermal combustion, without the addition of catalysts, requires very high temperatures—in the case of phenol, the combustion temperature reaches 720 °C—which is associated with very high financial outlays. Therefore, the method of catalytic combustion is more widely used—for comparison, the catalytic combustion temperature for phenol is 250 °C. The product of catalytic combustion of hydrocarbons and organic compounds containing oxygen is carbon dioxide and water, and in the case of improperly selected process parameters there is a risk of incomplete combustion and emission of toxic compounds (e.g., aldehydes). The role of catalysts is played by inorganic supports, such as silica, alumina, zeolite, and activated carbon, on which precious metals—platinum, palladium, copper, or vanadium—are deposited. The combustion of low concentrations of odors, about a few $mg/m^3$, is in most cases uneconomical, because all the heat needed to heat the gases must be supplied from external sources. In such cases, it is necessary to increase the odor concentration, by concentrating them in order to reduce costs [23]. For this purpose, a common practice is to combine combustion processes with adsorption [24]. First, the adsorbent is saturated with pollutants as a result of odor adsorption on active carbon, and then the pollutants are desorbed from the adsorbent and concentrated in the gas, which is then subjected to the combustion [5,22].

### 3.2. *Biological Methods of Odor Removal*

Biological methods of gas purification, based on the natural processes of decomposition of organic compounds occurring as a result of the metabolic activity of microorganisms, have gained an opinion in recent decades of the most beneficial methods of pollutant degradation. This opinion results from several significant advantages of biological methods: economy, ecological purity, lack of secondary pollutants, use of processes naturally occurring in nature, and high efficiency of pollution removal [20,25]. There are three main technologies used for air bio-purification: biofilters, bio-scrubbers and biotrickling filters. These methods differ in the type of layers and mobile phases as well as in the location of pollutant-degrading microorganisms [4,26,27].

#### 3.2.1. Biofilters

Deodorization by biofiltration, as shown in Figure 2, consists in passing a humidified, contaminated gas through a solid bed containing microorganisms capable of odors and VOCs degradation. The pollutants are sorbed and then absorbed by bacteria and decomposed into water and carbon dioxide [20]. In most cases, the biofilter bed consists of organic materials: wood bark, peat, straw, loosened soil, compost, coconut fiber, and activated carbon. The biofilter bed is piled up from one to several layers in such a way as to ensure contact of the entire gas stream with the bed and to maintain uniform aeration of the bed in order to prevent the growth of anaerobic bacteria, causing the bed to rot [19]. Bacteria, which have a natural ability to degrade odor and VOCs pollutants, form a biofilm on the surface of the biofilter bed and are selected according to the composition of pollutants present in the gas passing through the biofilter. In addition, they are provided with appropriate conditions for growth and development, e.g., by maintaining an appropriate pH in the bed and regularly supplying nutrients and mineral salts. Microorganisms are selected in such a way as to ensure their greatest diversity, which will enable the degradation of the widest range of pollutants [14]. Biofilters are a commonly used method of purifying gases from odors and VOCs emitted by the municipal sector because this method effectively removes both organic pollutants, including aromatic hydrocarbons (toluene and xylene), alcohols, aldehydes, organic acids, and amines as well as inorganic compounds such as hydrogen sulfide and ammonia. However, in the case of inorganic pollutants, it is necessary to control their concentration, due to the products formed as a result of their decomposition, which cause acidification of the biofilter environment. Moreover, the biofiltration method has some limitations on the concentration of pollutants in the treated gas. Biofilters are usually used for treatment of relatively large gas streams [4] (according to practice they are usually used for flows up to 5000 $m^3/h$), therefore they require a large mass exchange area and consequently a large size biofilter. The main disadvantages of biofiltration are the difficulty of controlling the process—maintaining the appropriate humidity and pH of the bed (which may become acidic)—as well as clumping of the filter material, relatively large installation size and lower treatment efficiency at high concentrations of pollutants. On the other hand, the advantages include low operating and investment costs, the possibility of purifying large streams of gases at low concentrations of pollutants [4,28].

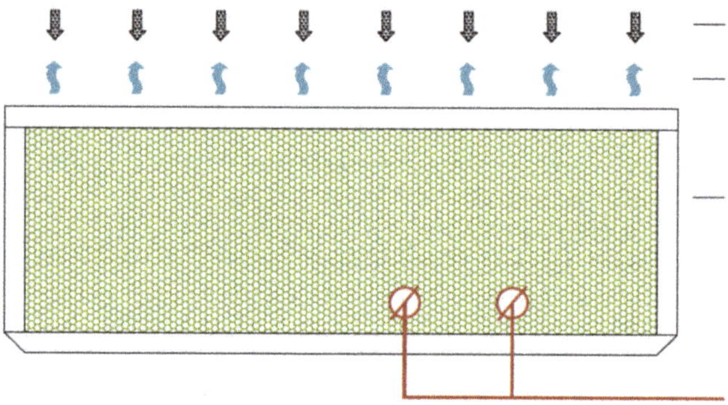


**Figure 2.** Diagram of a biofilter (own study, based on [5]).

### 3.2.2. Bioscrubbers

The principle of bioscrubber operation is based on two main stages that usually take place in separate devices [25]. In the first tank—the absorber—gaseous pollutants are absorbed into the liquid phase, which then goes to the second tank—the bioreactor. The bioreactor is filled with an aqueous suspension of microorganisms in which biodegradation of pollutants takes place. The liquid circulates through tanks supplied with air, nutrients for bacteria and pH adjusting solutions, while the excess of activated sludge is drained outside the system. The principle of operation and the structure of the bioscrubber are shown in Figure 3. The absorbers are filled with a bed that acts as a carrier for microorganisms. On the surface of the filling, microorganisms form a biofilm consisting of clumped bacterial cells and extra cellular polymeric substances (EPS) [29] capable of colonizing various environments and surfaces [30,31]. The undoubted advantage of bioscrubbers is the ability to control their operating parameters, such as pH, nutrient solution, aeration, which directly stabilizes their work. In addition, the installation is characterized by small dimensions, which is a significant advantage compared to biofilters, and there is no problem of clogging of the filter material. However, bioscrubbers generate large amounts of by-products such as excess sludge and contaminated, recirculated liquid; moreover, the operating costs of maintaining such an installation are much higher than in the case of biofilters [4,32].

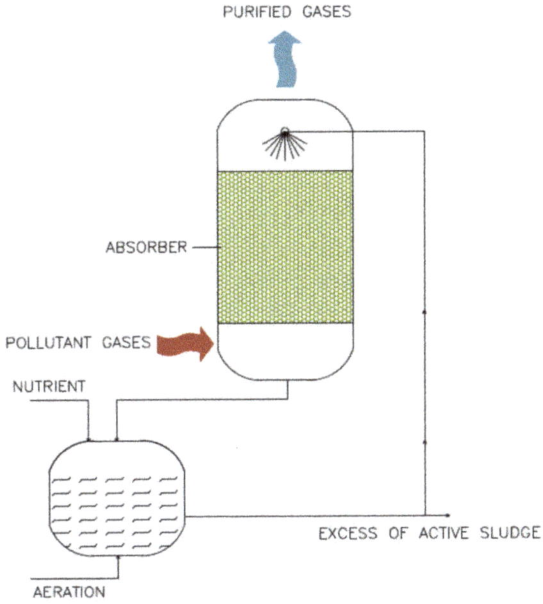


**Figure 3.** Diagram of a bioscrubber (own study, based on [5]).

Bioscrubbers are successfully used to remove odors, in particular $H_2S$, $NH_3$, and organic compounds with sulfur. However, due to their acidic nature, these substances cause a significant drop in pH, which may result in acidification of the medium circulating in the installation and a decrease in the efficiency of gas treatment [4].

3.2.3. Biotrickling Filters

In biotrickling filters, the process of absorption of pollutants into the liquid phase, and their biodegradation along with further liquid regeneration, takes place simultaneously in one tank [27]. The polluted gas enters the apparatus, in which it flows in the same direction or opposite to the liquid phase, in which the absorption of pollutants takes place. The liquid containing nutrients necessary for the development of microorganisms, along with the absorbed impurities, flows continuously as a thin film on the surface of the bed. As a result, the biofilm layer formed on the bed is constantly wetted, and biodegradation of pollutants to simple products such as water and carbon dioxide takes place there [4,27]. The liquid circulating in the plant is constantly recirculated, so there is no sludge waste. The scheme of biotrickling operation is shown in Figure 4. The bed in this type of installation is made of chemically inert materials, such as activated carbon, ceramic rings, glass balls, and plastic structures [20,27].

One of the advantages of biodegradation of gaseous pollutants in biotrickling filters over other methods is the ability to better control their operating conditions, such as maintaining an appropriate pH and composition of the medium circulating in the reactor. Moreover, the undoubted advantage is that the entire process is carried out in one tank, which saves a lot of space and total costs. On the other hand, the disadvantages that appear during the operation of biotrickling filters may be excessive growth of biomass inhabiting the bed, which may lead to the clogging of the bed and, as a result, a decrease in efficiency [33]. However, there are effective methods to counteract this, e.g., by temporarily increasing the flow of the liquid phase, which will result in breaking a part of the biofilm from the filling [34] or by appropriate selection of microorganisms eliminating excess bacterial biofilm (including protozoa), or by adding appropriate chemical to damage part of the bacterial biofilm [35].

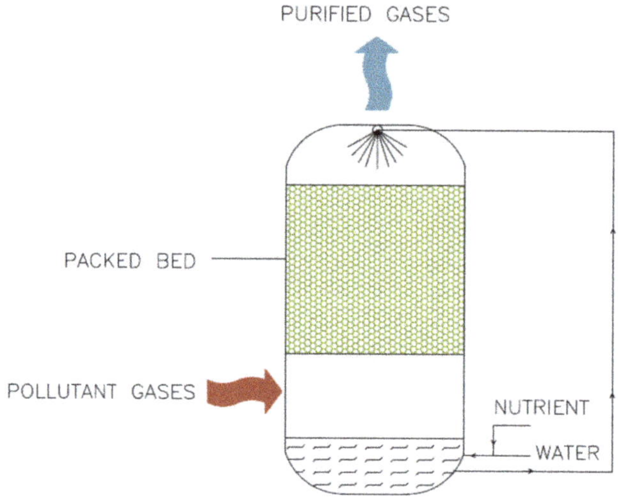


**Figure 4.** Diagram of a biotrickling filter (own study based on [36]).

## 4. Effectiveness of $H_2S$ and $NH_3$ Removal Using Biological Methods of Odor Degradation

Among the currently used odor removal methods, biological methods turn out to be the most attractive, in particular biofilters and biotrickling filters [7]. Among the biological methods of air purification, biofilters are relatively simple and the longest used methods; hence, also, the best known [14], there are many literature reports confirming the use of biofilters for odor removal.

### 4.1. Application of Biofilters to Remove $H_2S$ and $NH_3$

Chung et al. [37] studied the degradation of $H_2S$ and $NH_3$ using a biofilter. Impurities in the form of $H_2S$ and $NH_3$ were administered in various proportions. Their biodegradation efficiency was on the average level of over 95%, regardless of the $H_2S$ and $NH_3$ ratios used. The research was carried out in an experimental biofilter in the form of a column, on a laboratory scale. Moreover, it has been found that $H_2S$ can inhibit $NH_3$ removal, while $NH_3$ concentration has only a negligible effect on $H_2S$ removal.

Whereas Choi et al. [38] tested the $NH_3$ removal efficiency in two types of biofilters—with vertical and horizontal gas flow. Mixtures of organic materials such as compost, bark and peat were used as fillings, as well as inorganic material—pearlite (perlite). The result of the research was the determination of the ammonia removal capacity with the use of organic and inorganic media used in biofilters in order to select the most efficient filling. The organic packing achieved higher ammonia removal efficiency without significant pressure loss. When testing different types of gas flow, higher contamination removal efficiency was noted for horizontal gas flow reaching 100%.

Tymczyna et al. [39] also investigated the biodegradation efficiency of $NH_3$ with an open biofilter, but in this case the source of $NH_3$ was a poultry farm. The biofilter bed consisted of fibrous peat, coarse peat, wheat straw, wastewater treatment plant compost, and horse manure and was 1.2 m high, while the biofilter chamber area was 10 $m^2$. The efficiency of degradation of pollutants in the biofilter was tested in fivephases, in the initial phase of the experiment (after fivedays from filling the biofilter chamber) the efficiency was low—at the level of 36%, while after three months of biofiltration it increased to 89% and thus this result was the highest efficiency $NH_3$ removal during the experiment.

Pagans et al. [40] also investigated the effectiveness of $NH_3$ removal, this time from the gases emitted in the composting process, using a biofilter. The ammonia removal efficiency was nearly 96%. A significant decrease in the efficiency of $NH_3$ biodegradation was observed when its concentration at the inlet to the biofilter increased to over 2000 mg/m$^3$.

While Rehman et al. [41] investigated the performance of biofilters intended for $H_2S$ removal. The research was carried out in laboratory conditions, in six phases—starting with feeding only humidified air to the biofilter and gradually increasing the concentration of $H_2S$ with the subsequent phases. It was found that the biofilter most effectively removed $H_2S$ in the concentration range from 10 ppm to 30 ppm, then the efficiency was above 95%, while above these values the efficiency decreased, reaching an efficiency of 85% at an $H_2S$ concentration of 50 ppm.

In turn, the aim of the research by Omri et al. [42] was to investigate the degree of $H_2S$ removal in a biofilter filled with peat. The experiment was conducted on a pilot scale in a wastewater treatment plant in Tunisia. The concentration of $H_2S$ in the inlet gases ranged from 200 to 1300 mg/m$^3$, while the efficiency of $H_2S$ removal reached 99%.

Kavyashree et al. [43] investigated the use of a mixture of manure and rice husk as a filling in a biofilter to remove ammonia emitted by a municipal composting plant at concentrations of 500–700μg/m$^3$. The research was carried out with the use of a biofilter on a laboratory scale, for two variants of the bed depth: 20 cm and 40 cm. The effectiveness of $NH_3$ removal for a 20 cm bed depth was 61.5%, while for a 40 cm deep bed it was 71.45%. It was found that along with the increase in the number of bacteria in the deposit, the efficiency of ammonia degradation increases.

Aita et al. [44] investigated the effectiveness of removing $H_2S$ present in synthetic biogas using a biofilter filled with sawdust. The tests were carried out for 37 days, with an average $H_2S$ removal efficiency of 75 ± 13%, while the maximum efficiency was 97%.

Rabbani et al. [45] investigated the effectiveness of $H_2S$ and $NH_3$ removal from wastewater treatment plants, in a pilot-scale biofilter, under real conditions at the wastewater treatment plant. The experiment consisted of two stages, in the first stage, the biofilter was placed behind a chemical acid scrubber that removed $NH_3$ from gases. Thus, in the gases entering the biofilter, only $H_2S$ was present, which as a result of biological oxidation formed $H_2SO_4$, which was deposited at the bottom of the biofilter. The aim of stage I was to develop a sufficient amount of biofilm to remove $H_2S$ and to generate an appropriate amount of $H_2SO_4$ accumulated at the bottom of the biofilter to remove $NH_3$ in stage II. In turn, in the second stage of the experiment, gases containing a mixture of $H_2S$ and $NH_3$were introduced into the same biofilter, this stage lasted sevenweeks. The average $H_2S$ removal efficiency was 91.96% and $NH_3$ 100%. At the bottom of the biofilter, a small amount of effluent (0.2 mlof effluent/L reactor/day) accumulated in the form of ammonium sulfate. The authors noted that in the case of using biofilters on a full industrial scale, it would be necessary to look at the exact amounts of leachate produced.

Whereas the subject of research by Janas and Zawadzka [46] was the degradation of various odor compounds, including $H_2S$ and $NH_3$, emitted by the wastewater treatment plant with the use of a biofilter. The concentrations of $H_2S$ and $NH_3$ at the inlet to the biofilter were 154 μg/m$^3$ and 1799 μg/m$^3$, respectively, while their removal efficiency was 94% and 91%. However, despite the high efficiency of odor biodegradation, odor has not been completely eliminated.

Alinezhad et al. [6] compared the removal efficiency of odors consisting mainly of $H_2S$ and $NH_3$, emitted by a municipal wastewater treatment plant, using a chemical scrubber and a biofilter. The studies were conducted for 45 days. The biofilter was constantly fed with contaminated gas, while the efficiency of the removal of pollutants in the scrubber was tested only during those times of the day when odor concentrations were at the highest level. Both systems reported almost complete removal of $NH_3$, while the $H_2S$ removal efficiency was 95%. The experiment compared both methods in terms of technology and economy. The technological advantage of the chemical scrubber method over the biofilter was found due to the speed of gas loading and the limitations of the biofilter system. The degradation of both pollutants ($H_2S$ and $NH_3$) in a chemical scrubber was over 97%, while in the biofilter it was 92% for $H_2S$ and 99.5% for $NH_3$. However, in economic terms, the biological method of odor degradation in the biofilter turned out to be much more advantageous.

Baltrenas et al. [47] examined the effectiveness of air purification from ammonia in plate biofilters. The research was carried out with the use of different structures—a biofilter with straight lamella plates and a biofilter with wavy lamella plates. Various types of microorganisms were used, including yeast and bacteria. The efficiency of biopurification of air from ammonia was tested at various temperatures ranging from 24 to 32 °C. The best efficiency of ammonia biodegradation was achieved in a biofilter with wavy lamella plates and ranged from 84.2% to 87%.

Due to the simplicity of use and economic advantages for the recipient, biofilters have so far been the most frequently used method to removing odors, and thus the best known. However, for several decades, the odor removal technology in biotrickling filters has become an extremely competitive alternative. Examples of the use of biological degradation methods to remove $H_2S$ and $NH_3$ are shown in Table 4. Most likely, this is due to the legal restrictions on odor emissions and the need to find a method whose effectiveness reaches almost 100%, as well as the dynamic development of biotechnological methods of environmental cleaning in recent years.

### *4.2. Application of Biotrickling Filters to Remove $H_2S$ and $NH_3$*

The method of air purification using biotrickling filters has been successfully tested in various technological combinations for both leachate and gas purification (Table 4). Cox et al. [48] tested $H_2S$and VOC removal in a biotrickling filters on a pilot scale. Odor removal ($H_2S$) achieved an efficiency of 98%, but the simultaneous removal of VOCs achieved a much lower efficiency, which is influenced, among others, by drop in pH during $H_2S$ oxidation. Based on the pilot scale studies, it was concluded that the simultaneous removal of VOCs and odors ($H_2S$) is limited, which was not shown in previous laboratory scale studies [49]. Gabriel, Cox, and Deshusses [50] also investigated the removal of $H_2S$ emitted from wastewater treatment plants under real conditions on a full industrial scale. The results showed a high $H_2S$ removal efficiency despite the short gas contact time in the bioreactor caused by the high gas flows. These studies looked at only one compound—$H_2S$.

Aroca et al. [51] conducted experimental studies on $H_2S$ biodegradation using a laboratory scale biotrickling filter. They investigated the ability to remove $H_2S$ using two different bacterial strains (*Thiobacillusthioparus* and *Acidithiobacillusthiooxidans*), for different pH values and different concentrations of $H_2S$ in the inlet gas. The efficiency of $H_2S$ removal was compared for different concentrations at the inlet to the bioreactor and different contact times—better efficiency of $H_2S$ removal was noted—nearly 100%—for higher concentrations of $H_2S$ at the inlet to the reactor −4600 ppmv and 120 s residence time and 982 ppmv and 45 s residence time, than at the lower concentrations when the $H_2S$removal efficiency was 47%.

Ramirez et al. [52] also investigated the removal of $H_2S$ from gases in a Trickle Bed Bioreactor. The research was carried out in stable laboratory conditions on a bench-scale. The $H_2S$ removal efficiency was 98–99%.

Very broadly, Kasperczyk et al. described the use of Compact Trickle Bed Bioreactors to purify gases from VOCs and odors of various origins. Contaminated gases supplied to the reactor, which are the main source of carbon for bacteria, are absorbed into the liquid phase, and then diffuse into the bacterial biofilm inhabiting the reactor bed. In bacterial biofilm as a result of the metabolic activity of microorganisms, they are transformed into simple products such as water and carbon dioxide [20]. Nutrients needed by microorganisms for proper development are delivered in the form of a solution of mineral salts along with the liquid recirculated in the reactor, which constantly moistens the surface of the bed. An important advantage is the ability to control the conditions in the reactor, such as maintaining the appropriate pH, the composition of mineral salts, which ensure good conditions for the development of microorganisms, and temperature. Moreover, Compact Trickle Bed Bioreactors do not generate additional waste in the form of secondary pollutants, and are also a relatively inexpensive technology, which is conditioned by their operation at ambient temperature and atmospheric pressure [53]. Figure 5 shows a full-scale industrial Compact Trickle Bed Bioreactors.

**Figure 5.** Compact Trickle Bed Bioreactors in full industrial scale (Compact Trickle Bed Bioreactors, Manufacturer: Ekoinwentyka LTD, Ruda Śląska, Poland), Reproduced from [54], Industrial varnishing: 2020.

The latest published results of Kasperczyk et al. [20] presented the removal of VOC and $H_2S$ emitted by a sewage treatment plant with the use of a Compact Trickle Bed Bioreactor. The experiment was conducted on a semi-industrial scale, in a wastewater treatment plant. The $H_2S$removal efficiency at about 200 ppm concentration on inlet, was over 97%. During the experiment, jumps in $H_2S$ concentrations from 400 to 600 ppm were noted, which resulted in poisoning the bioreactor. However, after $H_2S$ concentrations were restored to normal, stable bioreactor operation was achieved within 3 h. Kasperczyk et al. [55] also investigated the biodegradation of a mixture of $H_2S$ and VOC from copper mines. The research was carried out in a Compact Trickle Bed Bioreactor, on a semi-industrial scale, in a copper mine, 1000 m underground. The bioreactor was filled with polyethylene rings. The efficiency of $H_2S$ removal was at the level of 80–99%—when the concentration of $H_2S$ was below 38 ppm, while when jumps in $H_2S$ concentrations of 40–60 ppm were noted, the efficiency of $H_2S$ removal decreased to 60–80%.

Sun et al. [56] examined a biotrickling purification filter for the treatment of $H_2S$ from a municipal wastewater treatment plant. In the research, the culture of microorganisms was excessive sludge, and the filling of the filter was made of polypropylene rings. It has been investigated that in the inoculums which was vaccinated with biotrickling filter there were such microorganism as *Pseudomonas* and *Thiobacillus*. The average $H_2S$ removal efficiency was 91.8%. In addition Sun et al. [57] also investigated the removal of hydrogen sulfide and volatile sulfur compounds using a two-stage biotrickling system containing acid- biotrickling filter and neutral- biotrickling filter. The contaminated gas came from wastewater treatment plant. Biotrickling filters was filled with polypropylene rings. The microorganisms most abundantly present in the biotrickling filter system were identified: *Acidithiobacillus* and *Metallibacterium*. The $H_2S$ biodegradation efficiency was 86.1%.

Chen et al. [58] tested the biodegradation efficiency of $H_2S$ in biotrickling filter in a pilot scale. The contaminated gas came from the sewage lift station. The biotrickling filter was filled with bamboo charcoal and inoculated with activated sludge from the wastewater treatment plant. During the research the removal rate was 99% with an inlet $H_2S$ concentration of 5–20 ppmv.

Most of the scientific reports analyzing the use of the method of biotrickling filters for odor removal concern the removal of only $H_2S$—considered to be the most persistent representative of odors. There are also many publications on the simultaneous removal of $H_2S$ and VOCs as components of odors. An equally persistent and harmful odor compound emitted by sewage treatment plants is ammonia $NH_3$.

Sakuma et al. [59] investigated the $NH_3$ removal from polluted air in a system consisting of a biotrickling filter, a denitrification reactor and a leachate treatment reactor (to prevent recycle of the effluent into the biotrickling filter). Composite balls made of ceramics and bovine bones were used as reactor packing. The biotrickling filter and denitrification reactor were inoculated with activated sludge from the wastewater treatment plant. $NH_3$ absorption and nitrification took place in the biotrickling filter, while nitrates and nitrites were removed in the denitrification bioreactor. Then the excess of dissolved COD and $NH_3$ was treated in the last reactor. $NH_3$was removed effectively, because in the first 15 days of operation the ammonia removal efficiency was 92–96%, while in the further stage of the experiment—after 21 days—the ammonia degradation efficiency did not drop below 96%, reaching 100% in several times.

While Moussavii et al. [60] investigated the removal of $NH_3$ in a biotrickling filter that developed a simultaneous nitrification/denitrification process. The bioreactor was filled with polyurethane foam, while the desired concentration of $NH_3$ flowing into the reactor was obtained by adjusting the air and $NH_3$ streams by trial and error. The results showed that this bioreactor would be able to completely remove 100 ppm $NH_3$ from the polluted gas with a 98.4% efficiency.

Huan et al. [61] investigated the efficiency of removing both $H_2S$ and $NH_3$ using a semi-pilot biological trickling filter reactor. As a filling of the biotrickling filter polyhedral spheres were used and it was inoculated with domesticated activated sludge. Microbiological analysis showed the presence of such microorganisms as *Dokdonella, Ferruginibacter, Nitrosomonas,* and *Thiobacillus*. The studies were conducted for 61 days and the removal rate of $H_2S$ was 98.25% and $NH_3$ was 88.55%.

Ying et al. [62] tested the ability of $H_2S$ and $NH_3$ biodegradation in a laboratory scale biotrickling filter, packed with porcelain Raschig rings and ceramsite. The maximum degree of $H_2S$ and $NH_3$ removal was over 99%.

Liu et al. [63] conducted research on integrated reactors in full-scale to determine the degree of odor removal (mainly $H_2S$ and $NH_3$), VOC and bioaerosols simultaneously. The polluted air used for the study came from the sludge dewatering room in wastewater treatment plant. The average biodegradation efficiency of the odors was 98.5%, with a flow rate of 5760 $m^3/h$, while the concentration of odors in the polluted air was recorded: $H_2S$ from 0.95 to 41.26 $mg/m^3$ and $NH_3$ from 0.91 to 21.37 $mg/m^3$.

**Table 4.** Examples of the use of biological degradation methods to remove $H_2S$ and $NH_3$.

| Type of Odor | Method of Biological Biodegradation | Type of Microorganism/Bacterial Strain | Parameters (Type of Filling, T, pH) | Efficiency | References |
|---|---|---|---|---|---|
| $H_2S$ | Biofilters | *Thiobacillus thioparus*($H_2S$), *Nitrosomonas europaea*($NH_3$) | 30 °C Ca-alginate beads | 95% | [37] |
| | | Sulfur Oxidizing Bacteria and microorganisms from compost | Compost pH = 7.5 | 95% | [41] |
| | | *Bacillus* sp., *Pseudomonas* sp., *Xanthomonoadacea* sp. | Peat | 99% | [42] |
| | | *Acidithiobacillus thiooxidans* | Wood chips | 75 ± 13% to 97% | [44] |
| | | Sulfur oxidizing bacteria | Acid resistant polyethylene packing material—AMB BiomediaBioballs | 91.96% | [45] |
| | | - | Pine bark | 94% | [46] |
| | | Activated sludge | Pieces of Poly Vinyl Chloride with compost | 84–99% | [6] |
| | Biotrickling filters | Raw influent water from plant (Hyperion treatment plant) | 7 layers of a PVC COOLdektm$^{tm}$Munsters | 98% | [48] |
| | | Heterotrophs, yeast, fungi, autotrophic sulfur-oxidizers | Pall rings, I biotrickling filter pH = 4.5, II biotrickling filter pH = 7 | ~100% | [49] |
| | | Primary and secondary sludge from Orange County Sanitation District | Polyurethane foam, T = 18–24 °C | ~98% | [50] |
| | | *Thiobacillus thioparus*, *Acidithiobacillus thiooxidans* | Volcanic stones, polypropylene rings, polyvinilclorure, pH = 5.5–7 | 100% | [51] |
| | | *Acidithiobacillus thiooxidans* | Polyurethane foam | 98–99% | [52] |
| | | *Pseudomonas fluorescens*, *Thiobacillus* sp. | Polyethylene rings, T = ~30 °C, pH = 5.5–7.5 | 97% | [20] |
| | | Bacterial strains | Polyethylene rings, pH = 5–7.5, T = ~30 °C | 80–99%; 60–80% | [55] |
| | | *Pseudomonas* sp., *Thiobacillus* sp. | Polypropylene rings | 91.8% | [56] |
| | | *Acidithiobacillus* sp., *Metallibacterium* sp. | Polypropylene rings | 86.1% | [57] |
| | | Activated sludge from Wastewater Treatment Plant (WWTP) | Bamboo charcoal | 99% | [58] |
| | | *Dokdonella* sp., *Ferruginibacter* sp., *Nitrosomonas* sp. and *Thiobacillus* sp. | Polyhedral spheres | 98.25% | [58] |
| | | *Acidithiobacillus* sp., *Thiobacillus* sp. | Raschig rings and ceramsite | 99% | [62] |

**Table 4.** *Cont.*

| Type of Odor | Method of Biological Biodegradation | Type of Microorganism/Bacterial Strain | Parameters (Type of Filling, T, pH) | Efficiency | References |
|---|---|---|---|---|---|
| $NH_3$ | Biofilters | *Thiobacillus thioparus*($H_2S$), *Nitrosomonas europaea*($NH_3$) | 30 °C Ca-alginate beads | 95% | [37] |
| | | Activated sludge from Wastewater Treatment Plant (WWTP) | Organic: compost, bark, peat Inorganic: pearlite | 100% | [38] |
| | | - | Fibrous peat, coarse peat, wheat straw, composts, horse manure | 89% | [39] |
| | | Compost | | ~96% | [40] |
| | | Nitrate oxidizing bacteria (*Nitrosomonas* sp., *Nitrobacter* sp.)—from cattle manure | Cattle manure, rice husk, gravel as a supporting media, 32–39 °C | 61.5%—for a bed 20 cm deep, 71.45%—for a bed 40 cm deep | [43] |
| | | - | Acid resistant polyethylene packing material—AMB BiomediaBioballs | 100% | [45] |
| | | - | Pine bark | 91% | [46] |
| | | Activated sludge | Pieces of Poly Vinyl Chloride with compost | 88–99.6% | [6] |
| | | Micromycetes: *Acremoniumstrictum*, *Aspergillus versicolor*, *Aureobasidium pullulans*, *Cladosporium* sp., *Penicillium* sp., *Gliocladiumviride*, *Stachybotrys* sp., *Cladosporiumherbarum*; Yeast: *Exophiala* sp., *Aureobasidiumpullulans*; Bacteria: *Rhodococcus* sp., *Bacillus subtilis* | Straight and wavy lamellar plates (hydrophilic synthetic texture), pH = 7, T = 24–32 °C | 84.2%–87% | [47] |
| | Biotricklingfilters | Acivated sludge from Wastewater Treatment Plant (WWTP) | Composite balls made of ceramics and bovine bones | 92–100% | [59] |
| | | Autotrophic and heterotrophic bacteria | Polyurethane foam | 98.4% | [60] |
| | | *Dokdonella* sp., *Ferruginibacter* sp., *Nitrosomonas* sp. *Thiobacillus* sp. | Polyhedral spheres | 88.55% | [61] |
| | | *Acidithiobacillus* sp., | Raschig rings and ceramsite | 99% | [62] |
| | | *Thiobacillus* sp., Ammonia Oxidizing Bacteria, Nitrite Oxidizing Bacteria | Activated carbon fiber | 98.5% | [63] |

## 5. Directions of Future Research

A review of recent research work underlines the need to use and implement modern, ecological, and cheap tools of biotechnology for odor removal into industrial practice. New physicochemical methods used for odor removal in wastewater treatment plants, such as ozone, UV rays, or non-thermal plasma, despite their high odor removal efficiency, are much more expensive than biological methods for the degradation of odors, and some of them generate emissions of secondary pollutants (including ozone) [23]. Higher operating costs of the above mentioned methods result, from the necessity to supply electricity and its high consumption. The non-thermal plasma method is used to degrade odors occurring in very small amounts (below 100 mg/m$^3$), when the concentrations of pollutants are higher, the increase in power of the device generates very high costs [23]. This technology causes also the formation of secondary pollutants, which in turn is associated with the need to combine at least two techniques of gas purification, and the resulting significant increase in financial expenditure [64,65]. Hołub et al. [64] achieved 90% odor reduction, but it was noted that not all compounds were removed—aldehydes and other hydrocarbons were removed to a small extent. However, the main advantage of this method is the small size of the installation [23].

The intensive development of modern ecological and innovative biotechnologies, and the steadily increasing amount of research over the last decade testify to the continuous development of this topic, and the focus of research on the possibility of potential implementation to full industrial scale. A review of the literature has shown that biological methods of odor removal give high effects of bio-purification of the air, up to 95–99%; moreover, their advantage is manifested primarily in the economic aspect as well as in terms of environmental friendliness. Therefore, it is necessary to develop and intensify processes based on biological methods of odor removal, in order to implement them to the full industrial scale. An example of the development of research on modern biotechnologies for odor removal is the Compact Trickle Bed Bioreactors, whose results and previous implementation indicate their potential versatility. Therefore, it is necessary to develop research in this area by testing the applicability of such technologies in various industries and in the municipal sector, e.g., wastewater treatment plants or landfills. Furthermore, the impact of extreme conditions and sudden changes of pollutant concentrations on the efficiency of air purification should be investigated, as most laboratory tests do not consider extreme overload conditions. It should be checked what parameters influence the inhibition or intensification of the efficiency of Compact Trickle Bed Bioreactors, e.g., what is the effect of a change in the composition of odors, which depends, among others, on atmospheric conditions and composition of wastewater delivered to the wastewater treatment plant, and how the activity and composition of microorganisms and their adaptation to the removal of variable concentration and composition of odors changes. Therefore, it seems important to study the influence of the parameters of the bio-treatment process and external conditions, in fact often deviating from stable laboratory conditions. In order to implement innovative biotechnological methods of odor removal into industrial practice, it would be necessary to carry out research in real conditions in industrial plants, waste management plants, municipal, and industrial wastewater treatment plants.

## 6. Summary

Among biological odor removal methods, the most commonly used so far isthe use of biofilters, which are effectivefor low concentrations of pollutants in the treated gas, can be used for large streams of polluted gases, are easy to build and operate, and are also relatively cheap. Currently, the method of gas purification in biotrickling filters turns out to be competitive to biofilters. It is a relatively new technology, whosegreat advantage, is the high efficiency of the biodegradation of pollutants, usually reaching 95–99%, and the ability to control the conditions in the reactor, such as maintaining an appropriate pH, mineral composition, which ensure good environment for the development of microorganisms. In addition, biotrickling filters do not generate additional waste in the form of secondary pollutants, and are a technology that does not generate furtheroperating costs, which means that there is no need to regenerate the bed, as is the case with biofilters, or to utilize excessive and harmful leachate, such as it is in the case of bioscrubbers. In view of these advantages, biotrickling filters are increasingly used in industry and are the subject of numerous studies.

The studies and results obtained so farshow that the research conducted on a laboratory scale does not reflect the actual conditions at a full industrial scale. Significantly higherflows in the gaseous phase, periodic changes, and sudden increases in pollutant concentrations, and the need to maintain an appropriate pH under such conditions are aspects that are not taken into account in laboratory tests because they are difficult to predict. Most of the odor and VOC removal tests performed so far, have been conducted on a laboratory, pilot or semi-industrial scale, while there are only few materials showing the implementation of biotrickling filter technology to a full industrial scale. The results show that the Compact Trickle Bed Bioreactor are a competitive method compared to other odor removal methods. Therefore it is necessary to strive to implement innovative biotechnologies to full industrial scale, which must be preceded by research carried out in real-life conditions at industrial plants and the municipal sector.

**Author Contributions:** Conceptualization, K.B. and D.K.; methodology, K.B., A.P.-G. and D.K.; validation, K.B. and D.K. formal analysis, K.B. investigation, A.P.-G., D.K.; resources, A.P.-G. and D.K.; writing—original draft preparation, A.P.-G.; writing—review and editing, K.B. and D.K.; visualization, K.B. and D.K.; supervision, K.B.; project administration, K.B. and D.K. All authors have read and agreed to the published version of the manuscript.

**Funding:** This research received no external funding.

**Institutional Review Board Statement:** Not applicable.

**Informed Consent Statement:** Not applicable.

**Data Availability Statement:** Not applicable.

**Conflicts of Interest:** The authors declare no conflict of interest.

## References


1. Seo, Y.-K.; Suvarapu, L.N.; Baek, S.-O. Characterization of Odorous Compounds (VOC and Carbonyl Compounds) in the Ambient Air of Yeosu and Gwangyang, Large Industrial Areas of South Korea. *Sci. World J.* **2014**, *2014*, 1–18. [CrossRef] [PubMed]
2. Michalak, A.; Krzeszowiak, J.; Pawlas, K. Does exposure to unpleasant odors harm human health? *Environ. Med.* **2014**, *17*, 76–81.
3. Abraham, M.H.; Gola, J.M.; Cometto-Muñiz, J.E.; Cain, W.S. The Correlation and Prediction of VOC Thresholds for Nasal Pungency, Eye Irritation and Odour in Humans. *Indoor Built Environ.* **2001**, *10*, 252–257. [CrossRef]
4. Mudliar, S.N.; Giri, B.; Padoley, K.; Satpute, D.; Dixit, R.; Bhatt, P.; Pandey, R.; Juwarkar, A.; Vaidya, A. Bioreactors for treatment of VOCs and odours—A review. *J. Environ. Manag.* **2010**, *91*, 1039–1054. [CrossRef]
5. Kośmider, J.; Mazur-Chrzanowska, B.; Wyszyński, B. *Odors*; Polish Scientific Publishers PWN: Odory, Poland, 2012.
6. Alinezhad, E.; Haghighi, M.; Rahmani, F.; Keshizadeh, H.; Abdi, M.; Naddafi, K. Technical and economic investigation of chemical scrubber and bio-filtration in removal of $H_2S$ and $NH_3$ from wastewater treatment plant. *J. Environ. Manag.* **2019**, *241*, 32–43. [CrossRef]
7. Liang, J.; Cheng, G.; Feng, H. Engineering practices of dezodorization for odor in urban sewage treatment plant in China. *Adv. Econ. Bus. Manag. Res.* **2016**, *30*, 86–91. [CrossRef]
8. National Research Council. *Acute Exposure Guideline Levels for Selected Airborne Chemicals*; The National Academies Press: Washington, DC, USA, 2008; Volume 6.
9. United States Department of Labor. Available online: https://www.osha.gov/SLTC/hydrogensulfide/hazards.html (accessed on 18 July 2020).
10. Nagaraj, A.; Sattler, M.L. Correlating Emissions with Time and Temperature to Predict Worst-Case Emissions from Open Liquid Area Sources. *J. Air Waste Manag. Assoc.* **2005**, *55*, 1077–1084. [CrossRef]
11. Sówka, I.; Sobczyński, P.; Miller, U. Impact of Seasonal Variation of Odour Emission from Passive Area Sources on Odour Impact Range of Selected WWTP. *Rocz. Ochr. Sr.* **2015**, *17*, 1339–1349.
12. Sobczyński, P.; Sówka, I.; Bezyk, Y. Characteristics of primary clarifiers odour emission variability and its influence on wastewater treatment plant odour impact range. *Interdiscip. IssuesEng. Environ. Prot.* **2015**, *6*, 356–363. (In Polish)
13. Adebayo, A.; Jekayinfa, S.; Linke, B. Effects of Organic Loading Rate on Biogas Yield in a Continuously Stirred Tank Reactor Experiment at Mesophilic Temperature. *Br. J. Appl. Sci. Technol.* **2015**, *11*, 1–9. [CrossRef]

14. Lewkowska, P.; Cieślik, B.M.; Dymerski, T.; Konieczka, P.; Namieśnik, J. Characteristics of odors emitted from municipal wastewater treatment plant and methods for their identification and deodorization techniques. *Environ. Res.* **2016**, *151*, 573–586. [CrossRef]
15. Barbusiński, K.; Kalemba, K. Use of Biological Methods for Removal of H2s from Biogas in Wastewater Treatment Plants—A Review. *Arch. Civ. Eng. Environ.* **2016**, *9*, 103–112. [CrossRef]
16. Zabava, B.S.; Voicu, G.; Ungureanu, N.; Dinca, M.; Paraschiv, G.; Munteanu, M.; Ferdes, M. Methods of biogas purification—A review. In Proceedings of the Acta Technica Corviniensis—Bulletin of Engineering, Hunedoara, Romania, 27 January 2019; Volume 12, pp. 65–68.
17. Okoro, O.V.; Sun, Z. Desulphurisation of Biogas: A Systematic Qualitative and Economic-Based Quantitative Review of Alternative Strategies. *Chem. Eng.* **2019**, *3*, 76. [CrossRef]
18. Awe, O.W.; Zhao, Y.; Nzihou, A.; Minh, D.P.; Lyczko, N. A Review of Biogas Utilisation, Purification and Upgrading Technologies. *Waste Biomass Valorization* **2017**, *8*, 267–283. [CrossRef]
19. Kwarciak-Kozłowska, A.; Bańka, B. Biofiltracja jako metoda unieszkodliwiania odorów powstających podczas kompostowania frakcji biodegradowalnej odpadów komunalnych i przemysłowych. *Inżynieria Ochr. Sr.* **2014**, *17*, 631–645.
20. Kasperczyk, D.; Urbaniec, K.; Barbusinski, K.; Rene, E.R.; Colmenares-Quintero, R.F. Application of a compact trickle-bed bioreactor for the removal of odor and volatile organic compounds emitted from a wastewater treatment plant. *J. Environ. Manag.* **2019**, *236*, 413–419. [CrossRef]
21. Alfonsín, C.; Lebrero, R.; Estrada, J.M.; Muñoz, R.; Kraakman, N.J.R.; Feijoo, G.; Feijoo, G. Selection of odour removal technologies in wastewater treatment plants: A guideline based on Life Cycle Assessment. *J. Environ. Manag.* **2015**, *149*, 77–84. [CrossRef]
22. Sarbak, Z. Bezpieczeństwoodoroweśrodowiska. PrzeglądNaukowo-Metodyczny. *Eduk. Bezpieczeństwa* **2014**, *7*, 151–171.
23. Wysocka, I.; Gębicki, J.; Namieśnik, J. Technologies for deodorization of malodorous gases. *Environ. Sci. Pollut. Res.* **2019**, *26*, 9409–9434. [CrossRef]
24. Kullavanijaya, E.; Trimm, D.L.; Cant, N. Adsocat: Adsorption/catalytic combustion for VOC and odour control. *Stud. Surf. Sci. Catal.* **2000**, *130*, 569–574. [CrossRef]
25. Gospodarek, M.; Rybarczyk, P.; Szulczyński, B.; Gębicki, J. Comparative Evaluation of Selected Biological Methods for the Removal of Hydrophilic and Hydrophobic Odorous VOCs from Air. *Processes* **2019**, *7*, 187. [CrossRef]
26. Barbusinski, K.; Kalemba, K.; Kasperczyk, D.; Urbaniec, K.; Kozik, V. Biological methods for odor treatment—A review. *J. Clean. Prod.* **2017**, *152*, 223–241. [CrossRef]
27. Rene, E.R.; Montes, M.; Veiga, M.C.; Kennes, C. Novel Bioreactors for Waste Gas Treatment. In *Environmental Chemistry for a Sustainable World*; Springer: Dordrecht, The Netherlands, 2012; Volume 2, pp. 121–164.
28. Chang, S.; Lu, C.; Huang, H.; Hsu, S. Removal of VOCs emitted from p-xylene liquid storage tanks by a full-scale compost biofilter. *Process. Saf. Environ. Prot.* **2015**, *93*, 218–226. [CrossRef]
29. Sutherland, I.W. Biofilm exopolysaccharides: A strong and sticky framework. *Microbiology* **2001**, *147*, 3–9. [CrossRef]
30. Kala, R.; Chauhan, H.; Rajput, A.; Kutty, R. Biofilm characterization and quorum quenching in pathogenic strains Staphylococcus aureus and Pseudomonas aeruginosa. *Int. J. Adv. Biotechnol. Res.* **2012**, *3*, 515–522.
31. Vu, B.; Chen, M.; Crawford, R.J.; Ivanova, E.P. Bacterial Extracellular Polysaccharides Involved in Biofilm Formation. *Molecules* **2009**, *14*, 2535–2554. [CrossRef]
32. Vergara-Fernández, A.; Quiroz, E.F.; Aroca, G.; Pulido, N.A.A. Biological treatment of contaminated air with toluene in an airlift reactor. *Electron. J. Biotechnol.* **2008**, *11*. [CrossRef]
33. Alonso, C.; Zhu, X.; Suidan, M.T.; Kim, B.R.; Kim, B.J. Mathematical Model of Biofiltration of VOCs: Effect of Nitrate Concentration and Backwashing. *J. Environ. Eng.* **2001**, *127*, 655–664. [CrossRef]
34. Smith, F.L.; Sorial, G.A.; Suidan, M.T.; Breen, A.W.; Biswas, P.; Brenner, R.C. Development of Two Biomass Control Strategies for Extended, Stable Operation of Highly Efficient Biofilters with High Toluene Loadings. *Environ. Sci. Technol.* **1996**, *30*, 1744–1751. [CrossRef]
35. Cox, H.H.; Deshusses, M.A. Biological waste air treatment in biotrickling filters. *Curr. Opin. Biotechnol.* **1998**, *9*, 256–262. [CrossRef]
36. Schiavon, M.; Ragazzi, M.; Rada, E.C.; Torretta, V. Air pollution control through biotrickling filters: A review considering operational aspects and expected performance. *Crit. Rev. Biotechnol.* **2016**, *36*, 1143–1155. [CrossRef] [PubMed]
37. Chung, Y.-C.; Huang, C.; Tseng, C.-P.; Pan, J.R. Biotreatment of $H_2S$- and $NH_3$-containing waste gases by co-immobilized cells biofilter. *Chemosphere* **2000**, *41*, 329–336. [CrossRef]
38. Choi, J.H.; Kim, Y.H.; Joo, D.J.; Choi, S.-J.; Ha, T.W.; Lee, D.H.; Park, I.H.; Jeong, Y.S. Removal of Ammonia by Biofilters: A Study with Flow-Modified System and Kinetics. *J. Air Waste Manag. Assoc.* **2003**, *53*, 92–101. [CrossRef] [PubMed]
39. Tymczyna, L.; Chmielowiec-Korzeniowska, A.; Saba, L. Biological treatment of laying house air with open biofilter use. *Pol. J. Environ. Stud.* **2004**, *13*, 425–428.
40. Pagans, E.; Font, X.; Sánchez, A. Biofiltration for ammonia removal from composting exhaust gases. *Chem. Eng. J.* **2005**, *113*, 105–110. [CrossRef]
41. Rehman, Z.U.; Farooqi, I.H.; Ayub, S. Performance of Biofilter for the Removal of Hydrogen SulphideOdour. *Int. J. Environ. Res.* **2009**, *3*, 537–544. [CrossRef]

42. Omri, I.; Aouidi, F.; Bouallagui, H.; Godon, J.-J.; Hamdi, M. Performance study of biofilter developed to treat $H_2S$ from wastewater odour. *Saudi J. Biol. Sci.* **2013**, *20*, 169–176. [CrossRef]
43. Kavyashree, B.A.; Ramya, N.; Sanjay Urs, A.M.; Chandan, K.B.; Shilpa, B.S.; Rashmi, M.B. Ammonia gas removal using biofilter. *Int. Adv. Res. J. Sci. Eng. Technol.* **2015**, *2*, 110–114. [CrossRef]
44. Aita, B.C.; Mayer, F.D.; Muratt, D.T.; Brondani, M.; Pujol, S.B.; Denardi, L.B.; Hoffmann, R.; Da Silveira, D.D. Biofiltration of $H_2S$-rich biogas using Acidithiobacillusthiooxidans. *Clean Technol. Environ. Policy* **2015**, *18*, 689–703. [CrossRef]
45. Rabbani, K.A.; Charles, W.; Kayaalp, A.; Cord-Ruwisch, R.; Ho, G. Pilot-scale biofilter for the simultaneous removal of hydrogen sulphide and ammonia at a wastewater treatment plant. *Biochem. Eng. J.* **2016**, *107*, 1–10. [CrossRef]
46. Janas, M.; Zawadzka, A. Biofiltration as an effective method for reduction of pollutant emissions. *Acta Innov.* **2017**, *23*, 43–50.
47. Baltrėnas, P.; Januševičius, T.; Zagorskis, A.; Baltrėnaitė-Gedienė, E. Removal of ammonia by biofilters with straight and wavy lamellar plates. *Int. J. Environ. Sci. Technol.* **2020**, 1–10. [CrossRef]
48. Cox, H.H.J.; Deshusses, M.A.; Converse, B.M.; Schroeder, E.D.; Iranpour, R. Odor and volatile organic compound treatment by biotrickling filters: Pilot-scale studies at hyperion treatment plant. *Water Environ. Res.* **2002**, *74*, 557–563. [CrossRef] [PubMed]
49. Cox, H. Co-treatment of $H_2S$ and toluene in a biotrickling filter. *Chem. Eng. J.* **2002**, *87*, 101–110. [CrossRef]
50. Gabriel, D.; Cox, H.H.J.; Deshusses, M.A. Conversion of Full-Scale Wet Scrubbers to Biotrickling Filters for $H_2S$ Control at Publicly Owned Treatment Works. *J. Environ. Eng.* **2004**, *130*, 1110–1117. [CrossRef]
51. Aroca, G.; Urrutia, H.; Núñez, D.; Oyarzún, P.; Arancibia, A.; Guerrero, K.; Oyarzún, P. Comparison on the removal of hydrogen sulfide in biotrickling filters inoculated with Thiobacillusthioparus and Acidithiobacillusthiooxidans. *Electron. J. Biotechnol.* **2007**, *10*, 514–520. [CrossRef]
52. Ramírez, M.; Gómez, J.M.; Cantero, D.; Paca, J.; Halecky, M.; Kozliak, E.I.; Sobotka, M. Hydrogen sulfide removal from air by Acidithiobacillusthiooxidans in a trickle bed reactor. *Folia Microbiol.* **2009**, *54*, 409–414. [CrossRef]
53. Kasperczyk, D.; Barbusiński, K.; Kozik, V. Use of Compact Trickle Bed Bioreactor for the Purification of Air from a Voc's Mixture—Preliminary Research. *Arch. Civ. Eng. Environ.* **2016**, *9*, 137–143. [CrossRef]
54. Kasperczyk, D.; Parzentna-Gabor, A.; Barbusiński, K.; Urbaniec, K. Ecoinnovationnot Only for the paint industry. *Industrialvarnishing* **2020**, *1*, 38–40. (In Polish)
55. Kasperczyk, D.; Urbaniec, K.; Barbusinski, K. Removal of Pollutants from the Air in a Copper-Ore Mine Using a Compact Trickle-Bed Bioreactor. *Chem. Eng. Trans.* **2014**, *39*, 1309–1314. [CrossRef]
56. Sun, S.H.; Jia, T.P.; Chen, K.Q.; Peng, Y.Z.; Zhang, L. Removal of hydrogen sulfide produced in a municipal WWTP using a biotrickling filter with polypropylene rings as the packing material and microbial community analysis. *Huan Jing KeXue* **2019**, *40*, 4585–4593. [CrossRef]
57. Sun, S.; Jia, T.; Chen, K.; Peng, Y.; Zhang, L. Simultaneous removal of hydrogen sulfide and volatile organic sulfur compounds in off-gas mixture from a wastewater treatment plant using a two-stage bio-trickling filter system. *Front. Environ. Sci. Eng.* **2019**, *13*, 60. [CrossRef]
58. Chen, Y.; Xie, L.; Cai, W.; Wu, J. Pilot-scale study using biotrickling filter to remove $H_2S$ from sewage lift station: Experiment and CFD simulation. *Biochem. Eng. J.* **2019**, *144*, 177–184. [CrossRef]
59. Sakuma, T.; Jinsiriwanit, S.; Hattori, T.; Deshusses, M.A. Removal of ammonia from contaminated air in a biotrickling filter—Denitrifying bioreactor combination system. *Water Res.* **2008**, *42*, 4507–4513. [CrossRef] [PubMed]
60. Moussavi, G.; Khavanin, A.; Sharifi, A. Ammonia removal from a waste air stream using a biotrickling filter packed with polyurethane foam through the SND process. *Bioresour. Technol.* **2011**, *102*, 2517–2522. [CrossRef]
61. Huan, C.; Fang, J.; Tong, X.; Zeng, Y.; Liu, Y.; Jiang, X.; Ji, G.; Xu, L.; Lyu, Q.; Yan, Z. Simultaneous elimination of $H_2S$ and $NH_3$ in a biotrickling filter packed with polyhedral spheres and best efficiency in compost deodorization. *J. Clean. Prod.* **2021**, *284*, 124708. [CrossRef]
62. Ying, S.; Kong, X.; Cai, Z.; Man, Z.; Xin, Y.; Liu, D. Interactions and microbial variations in a biotrickling filter treating low concentrations of hydrogen sulfide and ammonia. *Chemosphere* **2020**, *255*, 126931. [CrossRef]
63. Liu, J.; Kang, X.; Liu, X.; Yue, P.; Sun, J.; Lu, C. Simultaneous removal of bioaerosols, odors and volatile organic compounds from a wastewater treatment plant by a full-scale integrated reactor. *Process. Saf. Environ. Prot.* **2020**, *144*, 2–14. [CrossRef]
64. Holub, M.; Brandenburg, R.; Grosch, H.; Weinmann, S.; Hansel, B. Plasma Supported Odour Removal from Waste Air in Water Treatment Plants: An Industrial Case Study. *Aerosol Air Qual. Res.* **2014**, *14*, 697–707. [CrossRef]
65. Ren, B.; Zhao, Y.; Lyczko, N.; Nzihou, A. Current Status and Outlook of Odor Removal Technologies in Wastewater Treatment Plant. *Waste Biomass Valorization* **2018**, *10*, 1443–1458. [CrossRef]

*clean technologies*

*Review*

# Review of Methods for Assessing the Impact of WWTPs on the Natural Environment

**Joanna Bąk [1,*], Krzysztof Barbusiński [2] and Maciej Thomas [3]**

[1] Faculty of Environmental and Power Engineering, Cracow University of Technology, 24 Warszawska Street, 31-155 Cracow, Poland
[2] Department of Water and Wastewater Engineering, Silesian University of Technology, Konarskiego 18, 44-100 Gliwice, Poland; krzysztof.barbusinski@polsl.pl
[3] Chemiqua Water & Wastewater Company, 25/1 Skawińska Street, 31-066 Krakow, Poland; biuro@chemiqua.pl
* Correspondence: jbak@pk.edu.pl

**Abstract:** Environmental management in facilities such as wastewater treatment plants (WWTPs) allows for the implementation of the Deming cycle, and thus the constant improvement of the mitigation of the environmental impact. The correct diagnosis of the current state of functioning of the WWTPs, the identification of aspects that may have a measurable impact on the environment, and their assessment are of key importance. The article discusses the possible causes of the impact of WWTPs on the natural environment. Among other problems, such issues as energy consumption, noise and the formation of bioaerosols and odor nuisances were taken into account. Different ways of assessing the impact of wastewater treatment plants on the environment were collated, taking into account the need to assess not only the technological process itself but also the buildings during their use. The results of methods for assessing the environmental impact of wastewater treatment plants in selected countries were also compared.

**Keywords:** wastewater treatment plant; environment; impact; life-cycle assessment; environmental impact assessment; green building; environmental management system; environmental aspects; wastewater technology; management tool

**Citation:** Bąk, J.; Barbusiński, K.; Thomas, M. Review of Methods for Assessing the Impact of WWTPs on the Natural Environment. *Clean Technol.* **2021**, 3, 98–122. https://doi.org/10.3390/cleantechnol3010007

Received: 29 October 2020
Accepted: 20 January 2021
Published: 1 February 2021



## 1. Introduction

The natural environment is a good whose safety and proper protection should be taken care of in the course of social and economic development. In the era of dynamic technical and technological progress, this task is becoming more and more important. It is extremely important to take into account the environmental impact when planning any investment. For selected projects (which may have a significant or potentially significant impact on the environment), this is carried out by preparing an environmental impact assessment (EIA). Each country has appropriate procedures in this regard. In Poland, obligations in this respect are regulated by provisions [1,2]. Owners of various types of facilities can also apply for a green building certificate for new buildings. The BREEAM (Building Research Establishment Environmental Assessment Method) and LEED (Leadership in Energy and Environmental Design) are among the most popular multi-criteria assessment systems. However, the issue of environmental impact analysis should not be limited only to the design or construction stage. It should also apply to the process of using facilities and installations. In the case of buildings, the already mentioned green building certificates may be helpful, because some of the versions (called schemes) of these multi-criteria assessment systems are intended for already-existing buildings and require periodic renewal (BREEAM In-Use and LEED Existing Buildings Operation and Maintenance, LEED EB O+M). In turn, in the case of installation operation, both direct and indirect environmental management instruments are a solution. It is possible to mention here the necessity to obtain administrative permission (an emission permit or permit for operation and rationing) or a special type of environmental permission, i.e., an integrated permit. Apart from the mentioned legal and administrative instruments, economic instruments, such as fees for using the environment, should also be mentioned. However, these solutions apply only to some cases of economic undertakings. Moreover, they usually refer to only one type of environmental impact (except for installations requiring an integrated permit). Here, environmental management and the tools at its disposal play a role. The tools related to this management include implementation tools, such as environmental management systems (including the Eco-Management and Audit Scheme—EMAS—or systems according to ISO 14001 [3]) and diagnostic tools. Among the latter, the following deserve attention: life-cycle analysis (LCA) or the aforementioned environmental impact assessments (EIA), which can also be (voluntarily) carried out for existing facilities [4]. The implementation of the environmental management system in an enterprise is associated with the implementation of the Deming plan, and thus guarantees the continuous improvement of the effects of mitigating the impact of the organization on the environment, while the above-mentioned diagnostic tools make it possible to recognize the current state in terms of environmental impact and identify environmental aspects that have or may have an impact on the environment. This type of approach ensures greater care for the state of the environment and is not limited only to selected issues.

Facilities such as wastewater treatment plants are classified in Poland as projects that can always have a significant impact or can potentially have a significant impact on the environment; therefore, they require or may require an environmental impact assessment at the design stage (in accordance with Polish regulations [1,2]). This requirement does not apply to installations intended to serve a population of less than 400 inhabitants equivalent in accordance with the Act [5]. Wastewater treatment plants also require a water permit. Moreover, some wastewater treatment plants (WWTPs) require an integrated permit according to [6]. It should be added that fees are charged for pollutants discharged in wastewater into surface waters and soil. However, these requirements relate to the construction planning stage (i.e., EIA) and only to the discharge of wastewater to water or land (i.e., water permits and pollutant emission charges). Of course, certain standards must be met, and the waste generated during the technological process must be managed in accordance with specific regulations. However, the progressive technical and technological development enables the performance of the tasks of the wastewater treatment plant to be conducted in a way that is more environmentally friendly than before and consists of the use of more effective devices and energy obtained from sustainable sources. It is important that existing wastewater treatment plants can use new technologies and the latest technological achievements. However, to use them, knowledge about them is necessary, which can inspire people responsible for management to look for sources of financing for this type of investment. A holistic approach to the assessment of wastewater treatment plants is also important, i.e., covering not only technological processes, but also the impact of buildings on the environment during their operation.

The aim of the article was to review the methods of assessing wastewater treatment plants and analyze the environmental impact of wastewater treatment plants. As part of this task, an overview of the environmental aspects and possible environmental impacts of wastewater treatment plants was made. Selected tools and methods for assessing the impact of the wastewater treatment plant on the environment (resulting from both technological processes and building operation) are also discussed. Moreover, selected analysis of the environmental impact of planned or already-located WWTPs was performed. The implementation of the above-mentioned objective will allow for the comparison of various methods of assessing wastewater treatment plants and, at the same time, will enable decision-makers to more consciously choose a method adequate for current needs.

## 2. Review of Environmental Aspects and Impact on Environment of WWTPs

Each wastewater treatment plant is treated as an environmental protection facility due to its function. Nevertheless, it should also be remembered that it is an object that affects or may have a negative impact on the environment. The scale of this impact depends on many factors. The importance of the issue may be proved, inter alia, by the amount of treated wastewater or the degree of complexity of its treatment. Figure 1 shows the amount of treated municipal wastewater in Poland in 1998–2018, broken down by treatment method based on data from [7]. While the annual amount of wastewater fluctuates around the level of 1,200,000,000 m$^3$, one can clearly observe an increasing share of wastewater treated with increased nutrients removal. In 1998, only 24% of all treated municipal wastewater was treated with increased nutrients removal, while in 2018, it was already 85%. This shows the accompanying trend in the increase in the number of treatment plants with highly efficient unit processes, which in turn may translate into a potential increase in the impact of these facilities on the environment (ignoring the obvious benefits of a higher degree of wastewater treatment).

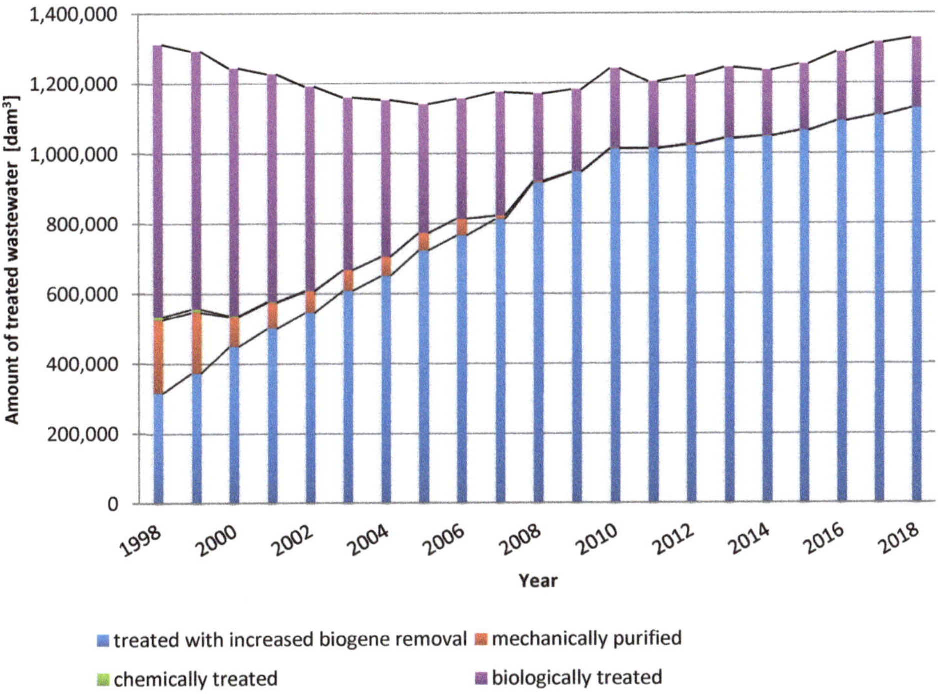


**Figure 1.** Change in the amount of municipal wastewater treated in Poland in 1998–2018, broken down by treatment method—own study based on data from [7].

The possible environmental aspects and the environmental impact of a wastewater treatment plant depend on several important factors. Among them, the most important are the type of treated wastewater; the location of the treatment plant, i.e., the distance from buildings and the type of receiver; the technology for the wastewater treatment and sewage sludge processing; the number of stages of wastewater treatment; the method of sewage sludge management; and the chemicals used (including coagulants and flocculants). Currently, solutions used in the field of minimizing energy consumption and the use of renewable energy sources are also becoming very important, because thanks to them, even to a minimal extent, it is possible to try to limit further climate change.

An environmental aspect is defined as an element of an organization's activities, products or services that interacts or can interact with the environment [3] or that has or can have an impact on the environment [8]. Below, an overview of the environmental aspects and environmental impacts typical of wastewater treatment plants is presented. At the beginning of the analysis, it is necessary to mention which components of the natural environment the WWTP has or may have an impact on. Figure 2 presents these elements of the environment, such as the air, surface and underground waters, soil, climate, landscape, fauna and flora, as well as human beings. Among the other elements that may be influenced and taken into account in some assessments, it is also worth mentioning material goods and monuments.

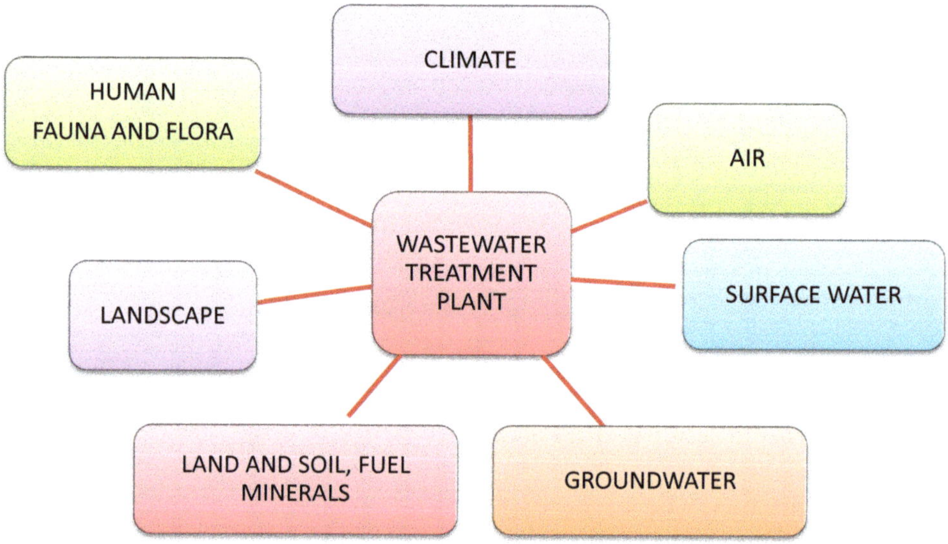


**Figure 2.** Environmental components influenced by the wastewater treatment plant during operation and/or failure.

One of the first problems associated with a wastewater treatment plant is the possibility of odors. The environmental aspect here is the release of odors, volatile organic compounds (VOCs) and bioaerosols from the conducted mechanical and biological treatment processes as well as sludge treatment processes. These bioaerosols contain microorganisms present in wastewater, while odors and VOCs cause discomfort to residents in the perception of the environment [9–11]. Bioaerosols settle on the surfaces of various elements (e.g., railings) or soil, or may be transferred with the wind to the surrounding areas. They spread about 800 m from the source of their formation [12]. Exposure to aerosols may pose a threat, first of all, to the employees of WWTPs and the population of nearby buildings [11,13], but also to animals [11]. They can also contaminate plants and surface waters [11]. It can be added that, also, the very presence of pathogenic microorganisms in wastewater and sludge is already an environmental aspect due to various possible routes of infection. Microorganisms can enter the human body through the respiratory tract, alimentary tract, mucous membranes or skin [11,14], and bioaerosol components are most often transported by air droplets or air dust [11]. The review of pathogens detected in wastewater treatment plants conducted by [14] mentions, among other things, viruses (e.g., enteroviruses, adenoviruses and rotaviruses), protozoa (e.g., *Giardia lamblia*), mold fungi (e.g., *Candida* spp.), mesophilic bacteria (e.g., *Pseudomonas* spp.), thermophilic bacteria (e.g., *Campylobacter* spp.), nematodes and tapeworms, and endotoxins. In addition to exposure to biological agents, there is also the threat of chemical substances. Among those to which workers of municipal wastewater treatment plants are exposed are heavy metals, volatile organic compounds, polychlorinated biphenyls, dioxins and polycyclic aromatic hydrocarbons [14].

Another environmental aspect related to the operation of wastewater treatment plants is noise generation. People (WWTP workers and the residents of nearby buildings) and animals can be susceptible to acoustic nuisance. The number of noise sources and the noise level depend, among other things, on the efficiency of the wastewater treatment plant, the applied technological cycle for wastewater treatment and the treatment of sewage sludge, the distribution of individual noise sources, and the technical advancement of the applied solutions. The noise at the wastewater treatment plant comes from working equipment, technological installations and means of transport. Noise is generated during mechanical and biological treatment processes as well as during sludge treatment. The acoustic nuisance is related, among other things, to the operation of pumps (including vacuum pumps), compressors, fans, centrifuges, means of transport for wastewater delivery, and the removal of screenings and sand [11,15,16].

Energy consumption is a very important environmental aspect related to the functioning of wastewater treatment plants. Energy is needed to carry out the treatment processes, transport of wastewater and sludge, and possible preparation for their reuse. Among the processes, energy is required for mechanical, biological and chemical treatment and disinfection. There is a demand mainly for electricity but also for gas and other fuels [17]. Energy consumption is associated with the depletion of non-renewable natural resources. Moreover, in the production of electricity in coal-fired power plants, pollutants are emitted to the atmosphere (including dust, sulfur dioxide, nitrogen oxides and carbon monoxide) [18,19]. These emissions also accompany vehicle traffic in the treatment plant, as well as transport related to the operation of the treatment plant (the delivery of wastewater, chemicals and waste disposal), and in this regard, the negative impact of this facility on the environment can also be noted (exhaust fumes from fuel combustion).

No less important an environmental aspect are the emissions of greenhouse gases from wastewater treatment plants. One of the greenhouse gases ($N_2O$) is mainly released from biological nitrogen removal processes in WWTPs with biological nutrient removal (BNR). The amount of $N_2O$ emissions from wastewater treatment plants is estimated at about 2.8–3% of the total emissions of this gas from all anthropogenic sources. Moreover, global $N_2O$ emissions from wastewater treatment were expected to increase by around 13% between 2005 and 2020 [20]. It is not the only emission, as municipal wastewater treatment plants with multistage activated sludge technology also generate methane ($CH_4$) and carbon dioxide ($CO_2$) [21]. Greenhouse gas emissions can come directly from the process units of a wastewater treatment plant, the effluent receiving environment, the biosolids receiving environment, as well as from greenhouse gas emissions related to the plant infrastructure, chemical consumption and operational energy consumption [22].

Taking into account the obvious benefits for the environment resulting from the operation of wastewater treatment plants, it should not be forgotten that when discharging treated wastewater into the environment (mainly surface waters), they also release substances into the environment (despite the compliance of the quality of treated wastewater with the established standards). This is especially true for pollutants such as suspended solids, organic compounds and biogenic compounds (in the case of municipal WWTPs). At industrial wastewater treatment plants, substances are also released into the environment, but their type depends strictly on the industry from which the wastewater comes.

Due to the fact that wastewater treatment plants are usually plants covering a large area of land and including cubature facilities, they have an impact on the landscape and the surface of the land (soil). Their construction reduces the biologically active surface. However, the impact on the landscape and the area of the land occupied must be analyzed at the stage of deciding on the location of the facility. At the operational stage, few actions to minimize such an impact can be taken.

Impacts on the land surface may occur in some sludge treatment or waste disposal processes in treatment plants (e.g., sludge plots and landfilling). On the other hand, in the case of the incineration of this waste and combustion of biogas resulting from sludge treatment, emissions to the atmosphere are recorded. This problem also applies to dewatered sludge from wastewater treatment plants at printed circuit board (PCB) manufacturers. PCB manufacturing requires photochemical processes, which use photopolymers dissolved in alkaline solutions ($Na_2CO_3$, $K_2CO_3$, NaOH, KOH, etc.). The spent alkaline developing solutions are pre-treated in on-site wastewater treatment plants. This process consists of the precipitation of the photopolymers in an acidic environment (pH 2–2.5, by using conc. $H_2SO_4$ or HCl), filtration, and dewatering with filter presses or filter bags. The resulting leachate requires further treatment (e.g., by using advanced oxidation processes, AOPs), and the sludge is then either processed further or disposed of. The large amounts of organic compounds present (mainly polymers) can be converted using combustion or carbonization processes [23–25].

The problem of the content of organic pollutants in dehydrated sludge also applies to those cases where only coagulation and flocculation processes are used for wastewater treatment. These processes only remove organic pollutants (without their decomposition) from the liquid phase and increase their concentration in the dewatered sludge [26,27].

Particular attention should be paid to certain industrial wastewater treatment plants that, in their wastewater treatment processes, produce only partially dewatered sludge, which contains high concentrations of heavy metals. Therefore, in the case that the sludge is not properly stored, transported and processed, it may cause harm to the natural environment. Sludge containing high concentrations of heavy metals is usually linked to conventional electroplating processes and also the production of printed circuit boards (PCBs). The raw wastewater from these industries often contains high concentrations of heavy metals (Cu, Ni, Sn, Zn, etc.). As a result of treatment processes in on-site wastewater treatment plants (e.g., by using chemical precipitation), heavy metal ions are precipitated in the form of insoluble sludge. These precipitates are separated from the liquid phase (i.e., treated wastewater) in the sedimentation process on lamellar settling tanks, and then dewatered with filter presses or with the use of filter bags. The use of different drainage methods results in varied water contents in the precipitates and, consequently, varied concentrations of heavy metals. In addition, the composition and physicochemical properties of sludge strongly correlate with the technological processes used in the manufacturing plants, the procedures used for water and wastewater management (rinsing processes and water recovery technologies), and the technical and technological processes in on-site wastewater treatment plants [28–30].

The environmental aspects of wastewater treatment plants should be considered not only for normal operation, but also in exceptional circumstances. In the event of a failure or disaster, there is a possibility of the contamination of surface water and groundwater as well as soil through leaks. This especially applies to failures of facilities filled with wastewater (pipelines and reactors, and tanks) and chemicals. It should be added, however, that such a negative impact on the environment is taken into account, and the treatment plants have appropriate safeguards. The implementation of pipeline monitoring can also be considered for the continuous assessment of their operation and control of possible leaks. The risk of potential impacts on individual components of the environment, i.e., the risk of the contamination of surface water, groundwater and soil, should be assessed, taking into account, inter alia, an analysis of the water–ground subsoil (the groundwater table level and soil types—permeable/impermeable) as well as the location of the treatment plant in relation to floodplains and exposure to earthquakes, and the quality of the proposed security measure design. The possible risk of such a negative impact depends, to a large extent, on these factors.

An often overlooked or underestimated issue when identifying the environmental aspects of wastewater treatment plants is the operation of office buildings, laboratories and social and technical facilities. These facilities are not without impact on the environment. They require energy needed for lighting, air conditioning, heating and the exploitation of room equipment (including computer equipment). During their operation, users also need hot and cold water and, possibly, gas. The amount of utilities used often depends on the quality of the building and its equipment. Another issue is the question of employees' commuting to work—the availability of, for example, bicycle paths or public transport. The above-mentioned issues are especially taken into account in a green building.

Figure 3 shows the more important environmental aspects of wastewater treatment plants. The diagram takes into account the effects on human beings, the air, surface water and the land surface and soil.

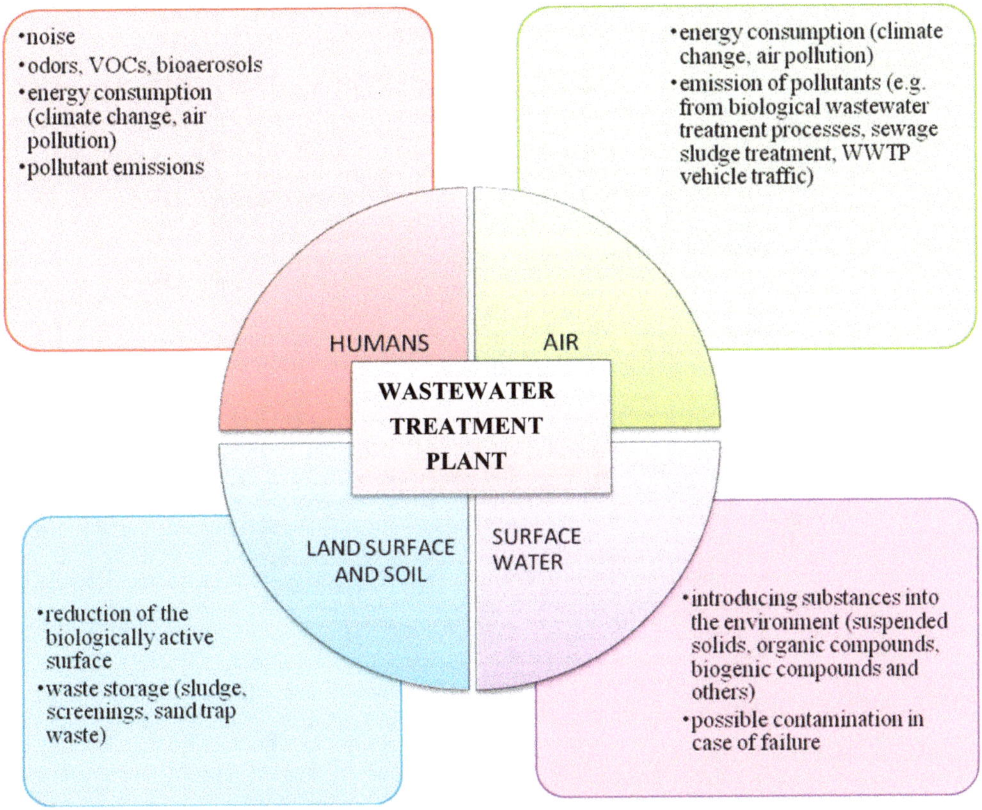


**Figure 3.** Main environmental aspects of wastewater treatment plants.

In summary, various types of processes are carried out at a wastewater treatment plant. These include mechanical treatment processes (screening, sedimentation and flotation), and chemical and biological treatment processes (including advanced nutrient-removal processes). Noteworthy are, among other things, nitrification (under aerobic conditions), denitrification (under anoxic conditions), biological dephosphatation (under anaerobic conditions) and chemical phosphorus precipitation. These processes lead to a reduction in, among other things, suspended solids, organic compounds and nutrients (nitrogen removal and phosphorus removal) in wastewater. As a result, the values of parameters such as biological oxygen demand and chemical oxygen demand are significantly reduced. An important aspect of the operation of a wastewater treatment plant is also the sludge treatment processes. Each of these processes has a different specificity and can also be carried out in a different way (e.g., by the activated sludge or bio-films in trickling filters). These processes generate different types of environmental impacts and with different scales of impact (different emissions, different compositions of waste, etc.), which depend on, inter alia, the composition of the wastewater. Thus, for mechanical treatment, the main impacts on the environment include the generated waste (screens and sand), noise, odors and VOCs, and for biological treatment, they include noise but also gas emissions to the atmosphere, and sludge with a need to be processed (treatment of sewage sludge). These processes are also associated with the emission of odors and VOCs. At this point, it is also worth adding that the issue of odor and VOC emissions during wastewater treatment is gaining importance, and there are more and more articles on this subject in the literature. Examples include [31,32]—on the methods of their removal or the assessment of the health hazards due to emissions of them. Moreover, practically all the processes in the treatment plant are carried out with energy consumption.

## 3. Possibilities for Assessing the Impact of Wastewater Treatment Plants on the Environment

Many wastewater treatment plants do not examine and evaluate the potential sources of nuisance and negative effects on the external environment, and risk assessments related to biological agents harmful to the health of employees are carried out sporadically on the areas of wastewater treatment plants [11]. However, in the era of technical and technological advancement with simultaneously progressing climate change, it would be highly recommended to take action in this area.

The impact of wastewater treatment plants on the environment can be considered at the design and operation stage. In the design phase, environmental impact assessment reports are prepared, while for the analysis of the environmental impact of a wastewater treatment plant in the operational phase, various tools are available for conducting such an assessment, although an environmental impact assessment (EIA) can also be used here [4]. A popular tool used for wastewater treatment plants is the life cycle assessment (LCA) technique. The analysis by this technique can be performed using different software and different methods. In the case of existing wastewater treatment plants, a useful tool may also be the identification of environmental aspects performed as part of the implementation of environmental management systems. The following is an overview of the methods used to assess wastewater treatment plants at different stages of their existence.

Among the ways of analyzing the impact of projects on the environment, a group of diagnostic tools can be distinguished, belonging to the group of environmental management tools, alongside implementation tools. This group includes, inter alia, environmental impact assessments, life-cycle analyses and environmental audits [4,33].

One of the most frequently used methods for assessing the environmental impact of wastewater treatment plants is the environmental impact assessment procedure. This method of evaluating projects was introduced in highly developed countries, including the United States, in the 1970s [34]. This assessment is performed at the design stage of this type of facility, and the procedure for carrying it out is regulated by the relevant regulations of a given country. The regulations of individual European countries in this area are related to the relevant European directives, i.a. [35–37]. According to the European Community Directive (the latest consolidated version) [35], now out of force, the scope of the EIA covered the direct and indirect impacts of the project on human beings, fauna and flora, soil, water, air, climate, landscape, and material assets and cultural heritage, and the interactions between these factors. In the European Union directive [36] currently in force, this provision has undergone some changes. Instead of effects on human beings, the effects on population and human health have been included, while the effects on flora and fauna have been replaced with effects on biodiversity. The effects on soil were supplemented with effects on land. Additionally noteworthy is adding the adjective "significant" before the word "effects". Currently, the scope of the impact assessment (according to the Directive [36]) should cover the direct and indirect significant effects of a project on factors such as population and human health, biodiversity, land, soil, water, air, climate, material assets, and cultural heritage and landscape, as well as the interactions between these factors. In addition, the impact on biodiversity should, in particular, take into account species and habitats protected under certain directives [37,38]. Among other, more important substantive EU environmental standards for wastewater treatment plants' projects, the following should be mentioned:

- Directive 2008/50/EC of the European Parliament and of the Council of 21 May 2008 on ambient air quality and cleaner air for Europe (establishes measures to assess air quality in the Member States on the basis of common methods and criteria) [39];
- Directive 2000/60/EC of the European Parliament and of the Council of 23 October 2000 establishing a framework for community action in the field of water policy (establishes a framework for the protection of inland surface waters, transitional waters, coastal waters and groundwater) [40];
- Directive 2008/105/EC of the European Parliament and of the Council of 16 December 2008 on environmental quality standards in the field of water policy (lays down environmental quality standards (EQS) for priority substances and certain other pollutants with the aim of achieving good surface water chemical status) [41];
- Directive 2006/118/EC of the European Parliament and of the Council of 12 December 2006 on the protection of groundwater against pollution and deterioration (establishes specific measures in order to prevent and control groundwater pollution) [42];
- Directive 2010/75/EU of the European Parliament and of the Council of 24 November 2010 on industrial emissions (integrated pollution prevention and control—lays down rules on the integrated prevention and control of pollution arising from industrial activities) [43];
- Directive 2008/98/EC of the European Parliament and of the Council of 19 November 2008 on waste (establishes the measures for protecting the environment and human health by, inter alia, preventing or diminishing the generation of waste, reducing the overall impacts of resource use, and improving the efficiency of such use) [44];
- Directive 2012/18/EU of the European Parliament and of the Council of 4 July 2012 on the control of major-accident hazards involving dangerous substances (sets up rules to prevent major accidents that involve dangerous substances, and limit their consequences for human health and the environment) [45];
- Council Directive of 21 May 1991 concerning urban wastewater treatment (91/271/EEC) —it concerns urban wastewater (collection, treatment and discharge) as well as wastewater from certain industrial sectors (treatment and discharge), and the objective of this act is protecting the environment from the adverse effects of these wastewater discharges [46];
- Council Directive of 12 June 1986 on the protection of the environment and, in particular, of the soil, when sewage sludge is used in agriculture (86/278/EEC)—the act aims to regulate the use of sewage sludge in agriculture in such a way as to prevent its harmful effects on soil, vegetation, animals and people, and thus encourage the proper use of such sewage sludge [47].

In the absence of a specific legal framework, EU soil protection policy is shaped by the EU Soil Thematic Strategy and the provisions of a range of other policy instruments, for example, the Environmental Liability Directive [48] and many others [49]. Current international conventions, in particular, those relating to nature protection, should also be taken into account. Environmental impact assessments carried out outside the European Union should be conducted in accordance with international (conventions), national and local regulations. For example, the procedure of Environmental Assessment for Wastewater System Improvements for the City of Sterling (Colorado) was performed in accordance with the Colorado Environmental Review Process and in conformance, inter alia, with the requirements of the National Environmental Protection Act (NEPA) [50].

Pursuant to Polish legislation [1], environmental impact assessment is a procedure for the environmental impact assessment of a planned project, including the verification of a project's environmental impact report, obtaining opinions and arrangements required by the act, and ensuring the possibility of public participation in the procedure. Such an assessment is carried out for objects that can always have a significant impact or can potentially have a significant impact on the environment. In Poland, according to [1], the environmental impact assessment for a project determines, analyzes and assesses the direct and indirect impacts of a given project on various factors (including the environment, population and landscape), the risk of major accidents, and the risks of natural and construction disasters, as well as the possibilities and methods of preventing and reducing the negative impact of the project on the environment and the required scope of monitoring.

Another environmental management tool worth mentioning, introduced at the turn of the 1960s and 1970s, is the ecological audit. The group of audits, also known as environmental reviews or ecological reviews, includes post-completion reviews (concerning, inter alia, controlling the effects of risk mitigation measures) and impact reviews (comparing the predicted environmental impacts in the EIA report with the actual impacts) [4].

Environmental management systems (EMSs), such as, for example, ISO 14001 or EMAS, are implementation tools in environmental management; however, as part of their implementation, the identification/determination of environmental aspects is performed. The identification of environmental aspects and impacts is a crucial part of any EMS [51]. According to the requirements of ISO 14001 [3], it is necessary to determine environmental aspects (of products, services and activities) taking into account current and planned activities and including aspects that can be controlled and influenced. The impact of the aspects and the importance of these aspects and effects should be assessed. The organization should also consider, inter alia, indirect aspects; past, present and future aspects; and actual and potential aspects [52]. In turn, in an EMS compliant with the regulation [8], each organization implementing this system must perform an environmental review of all the environmental aspects of the organization. According to the definition [8], it is an initial comprehensive analysis of the environmental aspects, environmental impacts and environmental performance related to the activities, products and services of an organization. Annex I of the Regulation [8] sets out its constituent parts. One of its elements is the identification of direct and indirect environmental aspects along with the identification of those that are significant. Among the methods useful in obtaining information needed to develop an environmental review is the life cycle assessment (LCA) technique [53]. According to [54], an initial environmental review should always be carried out before establishing and implementing an EMS to assess the organization's position towards the environment. There are different approaches to identifying environmental aspects and impacts. The ways to facilitate this task include the following methods: grouping, surveying, mass balancing, back calculating and Potpourri (combining the grouping methodology with the surveying methodology) [51]. An important element of environmental management systems in organizations is also carrying out audits (in accordance with the Deming plan), which contribute to activities aimed at minimizing the impact on the environment. It can therefore be concluded that the introduction of an environmental management system at a wastewater treatment plant is a way of not only assessing its impact on the environment (as part of identifying environmental aspects and impacts), but also permanently controlling it as part of the PDCA plan (Plan Do Check Act).

The life cycle assessment (LCA) is a rather popular diagnostic tool that allows for the analysis of the processes carried out in the wastewater treatment plant. The standard [55] defines life cycle assessment (LCA) as the compilation and evaluation of the inputs, outputs and potential environmental impacts of a product system throughout its life cycle. The rules of its conduct, requirements and guidelines are regulated by appropriate standards [55,56]. This tool allows for the identification, quantification and assessment of the potential impact and for determining a method of improving the quality of the environment [57]. According to [18], it is a recognized research method aimed at determining environmental hazards and, at the same time, allowing determining ways to improve the quality of the environment. It allows, for example, determining the impact of the designed/implemented technology on the environment throughout its life cycle [57]. According to [55,56], LCA studies consist of four phases, and the scopes of the studies depend on their subjects and purposes (intended uses of the studies). The LCA phases include defining the goal and scope of the study (phase 1), analyzing the set of "inputs" and "outputs" (phase 2), assessing the impact of the life cycle on the environment (LCIA) (phase 3) and interpreting the results (phase 4) [55,56]. The third phase of LCA research can be carried out with the use of various methods, often implemented into computer programs [57]. The method for carrying out an impact assessment (LCIA) is understood as a set of impact categories [58]. The methods differ in terms of categories but also in the parameters for characterizing the same categories, which sometimes leads to divergent results [57]. The following methods can be distinguished: Eco-Indicator 99, ReCiPe, CML (from the name of Centrum voor Milienkunde Leiden), IMPACT 2002+, EDIP, CED (Cumulative Energy Demand), Ecological Scarcity Method 2006, ILCD 2011 (International Reference Life Cycle Data System), TRACI (for the United States area), and USEtox [57–59]. Various programs are used to perform the LCIA phase. There are both commercial and free solutions. The programs used for this purpose include GaBi, SimaPro, TEAM, BEES, Umberto, ECO-IT, OpenLCA, and CMLCA, as well as Excel-based spreadsheets (MS Office package) or mathematical packages [59–62].

One of the most popular methods, Eco-Indicator 99, distinguishes three categories of damage—human health, ecosystem quality and resources [63]. Impact categories are assigned to each damage category. The human health category has the greatest number of impact categories. These are carcinogenic substances, organic and inorganic compounds that affect the respiratory system, ionizing radiation, ozone layer depletion and climate change. Ecotoxic substances, acidification and eutrophication as well as land management (land occupation and land conversion) have been assigned to the damage category "ecosystem quality", and to the resources category, the extraction of fossil fuels and mineral resources have been assigned [18,57,63]. The categories of damage refer to different units (e.g., human health can be expressed in units: DALYs, which means disability adjusted life years). Therefore, in the next step, normalization is used to maintain dimensionless degrees of importance. The final stage is the weighting process—performed by multiplying by appropriate importance factors [57]. The results are given in eco-indicator points (Pt), where 1 Pt represents one thousandth of the annual load of the environment, for one European inhabitant [18].

When assessing the environmental impact of wastewater treatment plants, the environmental impact of buildings during their operation should also be taken into account. Here, multi-criteria certification systems can be used. They can be seen as methods of assessing buildings and their impact on environmental, social and economic aspects [64]. Among the multi-criteria building assessment systems that can be applied to the buildings of wastewater treatment plants, those previously mentioned—LEED and BREEAM—can be distinguished. According to [65], LEED is the most widely used green building rating system in the world. This American system was introduced by the U.S. Green Building Council (USGBC) in 1998 as a pilot version (LEED 1.0), and in 2001, another version was launched—LEED 2.0 [65]. Version 4 was updated on 25 July 2019 with addenda [66], while currently, LEED version 4.1 from July 2020 applies [67]. The system is present in 160 countries and territories through participating projects [65]. Facility designs are assessed in nine main areas that address the key aspects of green building. These include integrative process, location and transport, sustainable sites, water efficiency, energy and the atmosphere, materials and resources, indoor environmental quality, innovation and regional priority [65]. Each of these categories includes specific subcategories—requirements, called credits—and points are awarded for meeting them. Their choice is free and depends on the investor and consultant [64]. Depending on the number of points scored, a grade is awarded. The facility can obtain rating levels from certified, through silver and gold, to the highest—platinum [65]. It is important that almost all the categories have specific critical requirements, the fulfillment of which determines obtaining a certificate [64]. It should be mentioned that there are different schemes depending on what facility is assessed. We can distinguish here the Building Design + Construction (BD+C) scheme for new projects, and the previously mentioned LEED EB O+M, but also Interior Design + Construction (ID+C) for complete interior design projects; LEED Homes, for residential buildings up to 6 stories; LEED Neighborhood Development, intended for new projects of land development or redevelopment projects; LEED Cities and Communities, for entire cities and the parts thereof; and LEED Zero, a certificate linked to the targets for reducing carbon dioxide emissions [68]. Among the mentioned systems for facilities such as wastewater treatment plants, the first-mentioned scheme can be mainly considered, as it is easier to meet these requirements at the design and construction stage of the facility.

Another very frequently used multi-criteria building assessment system is the BREEAM system. According to [69], this system is present in 87 countries, and almost 592,000 certificates have been issued to date. BREEAM certificates can be awarded to individual buildings, environments and infrastructure projects. There are a number of schemes and versions of this system depending, among other things, on the country, development phase and building function. There are technical standards for communities, infrastructure, new constructions, the in-use phase and the refurbishment and fit-out phase [69]. Another interesting solution is the "Bespoke" service offered by the system operator if it is not possible to find an appropriate technical standard for a given project. This requires, inter alia, adapting the criteria in the existing standards to a specific application [69]. The certification schemes are constantly evolving. From time to time, new and improved versions of the technical standards appear [70]. Among the latest versions of the schemes, the requirements for new structures in Great Britain from 2018 [71] can be distinguished. According to the source [69], categories such as energy, health and well-being, innovation, land use, materials, management, pollution, transport, waste and water are assessed during the certification process. Then, for each of these categories, the points obtained (called credits) are counted in relation to the achievable points. Each category is assigned a weight by which the obtained result for a given category is multiplied. The final grade is determined on the basis of the sum of the partial results. The building receives a grade depending on the results achieved. The field of grades ranges from acceptable (used only in the In-Use scheme), through pass, good, very good and excellent, up to the highest score—outstanding [69]. Besides an adequate number of points, the critical conditions and minimum requirements, which depend on the given certification level, must also be met [64]. Certificates that are issued for existing buildings (according to the BREEAM In-Use scheme) require periodical renewal. In turn, the certificates for new buildings are issued for an indefinite period (final certificates), but it should be noted that these are certificates issued in a specific version of a given scheme (that are modernized from time to time, as previously mentioned). As a result, in a sense, these final certificates may lose their validity and relevance after years. It is also possible to obtain a certificate at the design stage—a certificate called Interim [70].

In addition to the above-mentioned methods of assessing the impact of wastewater treatment plants on the environment, this group also includes articles, scientific papers and various types of expert opinions. They can be applied to virtually any aspect of WWTP functioning. The scope of the impact assessment is determined by the author or the customer and is usually related to the current problem at the wastewater treatment plant.

Table 1 presents the brief characteristics of various methods of assessing the environmental impact of an investment that can be applied to facilities such as wastewater treatment plants. The possibilities of applying a given assessment method (voluntary/obligatory and the type of an investment for which a given method is dedicated), the scope of the impact assessment and the forms of presenting the assessment are compared.

**Table 1.** Comparison of methods of assessing the impact of wastewater treatment plants on the environment.

| Assessment Method | Application | Scope of the Impact Assessment | Type of Assessment | Source |
|---|---|---|---|---|
| EIA (Environmental Impact Assessment) (based on Polish regulations) | Planned investments for which EIA is required (obligatory)<br>In the operational phase (voluntary) | The direct and indirect impact of the project on:<br>• The environment<br>• Population (health and living conditions)<br>• Material goods<br>• Monuments<br>• Landscape (including cultural landscape)<br>• Interactions between the above-mentioned elements<br>• Availability of mineral deposits<br>and in addition:<br>- Risk of major accidents as well as natural and construction disasters<br>- The potential for and ways of preventing and reducing the negative impact of the project on the environment<br>- The required scope of monitoring | Descriptive assessment in the form of an EIA report | [1,4] |
| LCA (Life Cycle Assessment) | Voluntary; for products (including services); carried out as part of the implementation of an EMS (Environmental Management System), or as part of scientific or other work for planned or existing plants | Depends on the subject and purpose of the study; it covers 4 phases: defining the purpose and scope of the study, analysis of the set of "inputs" and "outputs" (analysis of the inventory), assessment of the impact of the life cycle on the environment (LCIA) and interpretation of the results; for example, in the Eco-Indicator 99 method, there are 3 categories of damage and the corresponding impact categories (in brackets):<br>- Human health (carcinogenic substances, organic and inorganic compounds affecting the respiratory system, ionizing radiation, ozone layer depletion and climate change)<br>- Ecosystem quality (eco-toxic substances, acidification and eutrophication, and land management—land occupation and land conversion)<br>- Resources (extraction of fossil fuels and mineral resources) | Depending on the chosen method for the 3rd phase, e.g., in the Eco-Indicator 99 method, normalized and then weighted results are obtained in points for each damage category. Phase IV is the descriptive interpretation of the results. | [18,55–57,63] |
| EMS EMAS (Eco-Management and Audit Scheme) | Voluntary, existing organizations | Identification of all direct and indirect environmental aspects having a positive or negative impact on the environment with appropriate qualification and quantification; determination of significant aspects. | Descriptive assessment—environmental review combined with the assessment of environmental aspects through, for example, FLIPO forms (point assessment) and determination of significant environmental aspects; FLIPO (Flow—Legislation—Impact—Practices—Opinion) | [8,53] |
| EMS<br>ISO 14 001 | Voluntary, existing organizations | Identification of environmental aspects (products, services, activities) taking into account current and planned activities and including aspects that can both be controlled and influenced; assessing the impact of aspects and the significance of these aspects and effects; considering, inter alia, indirect aspects; past, present and future aspects; and actual and potential aspects. | Descriptive assessment—identification of environmental aspects | [3,51] |

**Table 1.** *Cont.*

| Assessment Method | Application | Scope of the Impact Assessment | Type of Assessment | Source |
|---|---|---|---|---|
| BREEAM (Building Research Establishment Environmental Assessment Method), various schemes | Voluntary; for:<br>- New construction<br>- Existing buildings<br>- Infrastructure<br>- Mixed use (bespoke projects)<br>but also available for:<br>- Refurbishment and fitting out<br>- Communities | Several dozen sub-categories are assessed, grouped into the following categories: energy, health and well-being, innovation, land use, materials, management, pollution, transport, waste and water. | A certified rating (acceptable—only for In -Use scheme—pass, good, very good, excellent, outstanding) and percentage score and certificate with star rating | [64,69] |
| LEED (Leadership in Energy and Environmental Design ) (v4) | Voluntary; for:<br>- New buildings (Building Design and Construction)<br>- Existing buildings (Building Operations and Maintenance)<br>but also includes other types of projects (interior design and construction, neighborhood development, cities and communities, homes) and LEED Zero | Compliance with the creditsgrouped in 9 areas isassessed:<br>- Integrative process<br>- Location and transport<br>- Sustainable sites<br>- Water efficiency<br>- Energy and atmosphere<br>- Materials and resources<br>- Quality of the internal environment<br>- Innovations<br>- Regional priority | Number of points and the corresponding rating level (certified, silver, gold, platinum) and certificate | [64,65] |
| Other—scientific articles and studies, expert opinions | Voluntary | Depending on the authors or the customer, usually related to specific problems occurring in the facility | Descriptive assessment | i.a.: [72–74] |

## 4. Review of Environmental Impact Analyses of Wastewater Treatment Plants Conducted in Various Parts of the World, with Particular Emphasis on Poland

The authors, based on the available literature, selected analyses of the impacts of wastewater treatment plants on the environment to review. Both analyses performed at the investment planning stage or its expansion and modernization, and assessments carried out during operation, were taken into account. Examples of assessments of wastewater treatment plants performed with different methods, which are discussed in the previous chapter, were selected. Particular attention was paid to examples from Poland, due to the availability of information in this regard.

The environmental impact assessment (EIA) of a wastewater treatment plant is one of the most frequently used methods of the assessment of the impact of a wastewater treatment plant on the environment, which is related to its obligatoriness. They most often concern the construction of a wastewater treatment plant or its extension or modernization. Both over the years and across countries, differences in the approaches to environmental impact assessments and their structure can be observed, although they basically all contain the same elements. Depending on the region of the world, attention is paid to other issues related to local conditions. For example, in the environmental impact assessment of a wastewater treatment plant located in the Port Said region [75], in the identification of possible environmental impacts, the possibility of mosquito breeding and the risk of disease transmission associated with it (mainly malaria) were taken into account. This part of the study also highlighted the (positive) impact of the treatment plants on bathing water quality and tourism. The environmental assessment of the wastewater system improvements for the County of Logan in the United States (including the modernization of the wastewater treatment plant) took into account the possible impact of the project on the wetlands in these areas, as well as on floodplains and their management [50]. In turn, the impact assessment of the reconstruction of the wastewater treatment plant for Minsk (Belarus) included, in addition to the part on environmental impacts, a part on social impact [76]. There are also differences in the level of detail of impact assessments, partly related to the scope of the assessed investment and its size. There are reports that are very detailed and those prepared in a brief and laconic manner. Depending on the scope of the investment, there may be impact assessments for improvements to the wastewater system e.g., [50], the wastewater treatment plant only e.g., [75,77,78] or the stage of its expansion or modernization e.g., [79,80].

Due to the importance of environmental impact assessment procedures, work has also been conducted to analyze or evaluate these impact assessments, as well as to propose necessary changes. The analysis presented in [81] of 11 selected Polish reports on the impact of wastewater treatment plants on the environment shows that these elaborations, assessed according to the proposed criteria, are far from complete descriptions of the variables, and the conclusions are too general for carrying out an objective decision-making process. This study [81] shows that the authors of the reports were the least in depth when describing and analyzing cumulative impacts, threats to groundwater and losses of material goods. However, the impact of the wastewater treatment plant on changes in the acoustic climate, air quality and surface waters was assessed in detail. It should be added that the impact on the acoustic climate, according to the study [81], was analyzed by the authors in the most comprehensive way (in relation to other variables). In the work [82], 33 processes of the environmental impact assessment of wastewater treatment plants in Spain were reviewed through records of decision (RODs). The most frequently identified impacts during the operation of a WWTP were odors from the depuration process and the sludge treatment, noise from pumps and the visual impact of the facilities. The negative impact on water quality was also mentioned. It was indicated that in some cases, an amount of total nitrogen that causes eutrophication could be generated.

Article [81] points to the need for creating facilitating tools for decision-makers to help them to make choices, but with the assumption that these instruments will constitute a reliable source of knowledge. Due to the need for comprehensiveness when deciding on environmental projects such as wastewater treatment plants, and the need to take into account environmental, sociopolitical and economic factors, in addition to mandatory environmental impact assessments, the paper [83] proposes the use of the decision support concept. The developed concept based on multi-criteria methods (analytic hierarchy process, AHP, and preference ranking organization method for enrichment evaluations, PROMETHEE) was used on a specific case study—choosing the location of a sewage treatment plant for the city of Kutina in Croatia.

The LCA technique can also be considered as a decision support tool in the field of the environmental improvement of an operated wastewater treatment plant, an example of which is the work [84]. The currently conducted technological processes as well as the management of byproducts (sludge and biogas) can be assessed and compared in this way [84–87]. Typically, the LCA technique is used when considering possible options for modernizing an existing wastewater treatment plant. The studies [84,88] can serve as an example here. In [84], the current situation of a wastewater treatment plant was assessed, and alternatives for improvement were identified, while in [88], the LCA technique made it possible to compare the environmental performance of five wastewater and sludge management scenarios in a WWTP in Italy. This technique can also be used when selecting a technology variant for a planned wastewater treatment plant. In Poland, the LCA technique is used to assess the production of alternative flocculants and the treatment of industrial wastewater with their use. The paper [89] describes the use of the LCA technique to identify the sources and assess the impact on the environment of the stage of the potential production of new-generation flocculants synthesized from post-production polystyrene waste and the stage of wastewater treatment with the use of synthesized products. In turn, in [90], the LCA technique was used to evaluate the impact of a new flocculant used in the treatment of wastewater from the metallurgical industry, taking into account the impact of the flocculant production process on the environment. It presents the results of an analysis carried out with the LCA technique for the appraisal of the environmental impact of the modified waste phenol–formaldehyde resin (which is called Novolak) [90].

It should be added that apart from the various types and scopes of assessments using this technique, there was also a proposal of a new methodology for conducting LCA for wastewater treatment plants in [91]. Its purpose is to avoid limitations in the interpretation of LCA results. It is based on an evaluation of the net environmental benefits (NEB) and requires an assessment of the potential impacts of wastewater discharge without and after treatment, in addition to the life-cycle impact assessment of the wastewater treatment plant [91].

The way to conduct regular assessments, but for existing wastewater treatment plants, is to implement an environmental management system. An example of this may be the implementation of the ISO 14001 system. It is also observed that many entities decide to integrate the environmental management system with another management system (for quality and/or safety), thus creating an integrated management system. ISO 14001 certificates are issued for wastewater treatment plants, an example of which is the "Klimzowiec" wastewater treatment plant in Poland [92], or for water supply and sewage companies, usually covering the production and supply of water as well as the collection, treatment and discharge of wastewater, e.g., [93–95]. Another example is the EMAS scheme. This management system is not yet very popular in Poland, as evidenced by the number of registered organizations, but several wastewater treatment plants carry out this method of assessing their environmental impact. According to the register [96], the Polish wastewater treatment plants implementing the EMAS environmental management system include the final wastewater treatment plant of PGE GiEK S.A. in Brzezie near Opole, wastewater treatment plants managed by the Water Supply and Sewage System Company of the Częstochowa District (joint-stock), wastewater treatment plants in Suszec and in Ornontowice managed by the Enterprise of Water Management and Reclamation (joint-stock) in Jastrzębie Zdrój, the wastewater treatment plant in Tychy, and the "Hajdów" wastewater treatment plant in Lublin. Due to the fact that the EMAS system is, in simple terms, an extension of the ISO 14001 system, it is possible for a wastewater treatment plant to have more than one environmental management system. This is the case for the "Hajdów" wastewater treatment plant in Lublin. The fact that WWTPs have the ISO 14001 certificate or have been entered into the EMAS register proves that the environmental aspects of the wastewater treatment plant have been identified at least once. On the other hand, maintaining and perfecting the environmental management system allows for the constant monitoring of the impact of a WWTP on the environment. For example, in the Water Supply and Sewage System Company of the Częstochowa District (joint-stock), which has an implemented EMAS system, in the scope of wastewater collection and treatment, the following environmental performance indicators are checked: energy efficiency, material efficiency (flocculants, coagulants and chlorinated lime), water consumption, the mass of generated waste (total and hazardous) and emissions [97]. Moreover, by comparing the situation before the implementation of the system and after 8 years of its operation, a significant reduction in the amount of waste from wastewater treatment processes that is subjected to landfilling was observed [97].

In a holistic approach, to assess any environmental impacts of a wastewater treatment plant, it is also necessary to consider buildings and their impacts. For this purpose, multicriteria evaluation systems are used, e.g., BREEAM or LEED. There is no wastewater treatment plant in Poland with buildings certified by the BREEAM or LEED system. Only a few wastewater treatment plants in the world are certified with these systems, but important is that such facilities do exist. A project worth mentioning is the new Aéris wastewater treatment plant for Cagnes sur Mer in France. At the beginning of 2018, it obtained the BREEAM Interim certificate in the International 2013 New Construction scheme for industrial projects, achieving a result of 56.5% and a very good rating [98]. The treatment plant is located between a railway line and a motorway to minimize its impact on residents and the environment. According to the plans, in the event of heavy rainfall, the treatment plant will treat the wastewater produced by the equivalent of 160,000 inhabitants. It is called a net positive energy wastewater treatment plant due to the fact that, ultimately, the plant will use less energy than it produces [99].

There are wastewater treatment plants, their selected buildings or buildings of water and sewage companies certified with the LEED system or under certification, as evidenced by the LEED project databases [100,101]. Among the facilities of the LEED certified WWTPs, the wastewater treatment plant in Dryden, Canada, deserves special attention. According to Dryden's Public Works Manager the Dryden Wastewater Treatment Plant is a LEED–certified structure which covers the building, and also the processes and the plant [102,103]. While other facilities have buildings that have obtained LEED certificates, the designers of the company responsible for the design of the Dryden plant believe it is the first entire WWTP in North America to achieve LEED certification [103]. One of the factors contributing to the certification is an in-floor heating system, which takes heat from the treated wastewater before it is discharged [102]. Other solutions that deserved credit -towards LEED include reusing treated wastewater to wash down and clean the facility, reusing heat generated by blowers, minimizing light pollution, having storage for bicycles, having showers for the crew, promoting green modes of transport, and diminishing water consumption through the use of water fittings such as waterless urinals or low-flow showers, sinks and toilets [103].

Table 2 presents basic information on selected certified facilities related to wastewater treatment plants in the United States and Canada. Data such as the scheme in which they were assessed, the year of issuing the certificate, the number of points (scored and the maximum possible) and the requirements for which individual objects obtained points in selected categories were compared. Selected areas, especially those associated with the impact on the environment, are included, i.e., sustainable sites, water efficiency, and energy and the atmosphere. In addition, points earned in the innovation category were also included. Credits such as those for indoor air quality and materials and resources were omitted. When analyzing the data contained in the table, it can be observed that the buildings of the wastewater treatment plant have received different ratings, but these are rather lower ratings (certified or silver level). It should be added that the mere obtaining of the certificate proves that all the critical requirements specified in the individual categories are met. It can be seen that the wastewater treatment plants in the United States scored points on different requirements, while the Canadian facilities obtained points on almost the same credits. Almost all the WWTPs received points for choosing a location, alternative transport and reducing water consumption. A frequently fulfilled requirement (but to different extents, i.e., with different numbers of points scored) was that to optimize energy efficiency.

**Table 2.** Data on LEED certification for selected facilities related to wastewater treatment plants.

| Project | Scheme | Score, Certification Level, Year of Certification | Selected Credits with Points Awarded | Source |
| --- | --- | --- | --- | --- |
| Sanford Wastewater Treatment Plant, Sanford, United States | LEED BD+C, New Construction, v2 LEED 2.2 | 38/69 points, silver, 2015 | - Sustainable sites (site selection, alternative transportation, stormwater design—quantity control)<br>- Water efficiency (water-use reduction, water-efficient landscaping)<br>- Energy and atmosphere (optimize energy performance, enhanced refrigerant management)<br>- Innovation (innovation in design) | [100] |

**Table 2.** *Cont.*

| Project | Scheme | Score, Certification Level, Year of Certification | Selected Credits with Points Awarded | Source |
|---|---|---|---|---|
| Central Wastewater Treatment Plant Maint, Dallas, United States | LEED BD+C, New Construction, v2 LEED 2.1 | 27/69, certified, 2009 | - Sustainable sites (site selection, alternative transportation, reduced site disturbance—development footprint and heat island effect—roof and non-roof, light pollution reduction)<br>- Water efficiency (water-efficient landscaping)<br>- Energy and atmosphere (optimize energy performance, additional commissioning)<br>- Innovation (innovation in design) | [100] |
| Triangle Wastewater Treatment Plant, Durham, United States | LEED BD+C, New Construction, v2 LEED 2.0 | 27 points, certified, 2005 | No data available | [100] |
| Florence Regional Wastewater Management Facility, Florence, United States | LEED BD+C, New Construction, v2 LEED 2.2 | 39/69, gold, 2013 | - Sustainable sites (site selection, alternative transportation, site development—maximize open space, heat island effect—roof)<br>- Water efficiency (water-use reduction, water-efficient landscaping)<br>- Energy and atmosphere (optimize energy performance)<br>- Innovation (innovation in design) | [100] |
| Control Building for Sewage Treatment Plant | LEED BD+C: New Construction v2—LEED 2.2 | 36/69, silver, 2011 | - Sustainable sites (site selection, alternative transportation, site development—maximize open space, stormwater design—quality control, heat island effect—roof, light pollution reduction)<br>- Water efficiency (water-use reduction, water-efficient landscaping)<br>- Energy and atmosphere (optimize energy performance, enhanced refrigerant management, green power)<br>- Innovation (innovation in design) | [100] |
| City of Dryden Wastewater Treatment Plant, Dryden, Ontario, Canada | LEED Canada NC 1.0 | 33/70, silver, 2016 | - Sustainable sites (site selection, alternative transportation, heat island effect—roof, light pollution reduction)<br>- Water efficiency (water-use reduction, water-efficient landscaping, innovative wastewater technologies)<br>- Energy and atmosphere (optimize energy performance, ozone protection, green power)<br>- Innovation (innovation in design) | [101] |
| Wood Buffalo Wastewater Treatment Facility, Fort McMurray, Alberta Canada | LEED Canada NC 1.0 | 29/70, certified, 2012 | - Sustainable sites (alternative transportation, stormwater management—rate and quantity, treatment)<br>- Water efficiency (water-use reduction, water-efficient landscaping, innovative wastewater technologies)<br>- Energy and atmosphere (optimize energy performance, ozone protection, green power)<br>- Innovation (innovation in design) | [101] |

In addition to the mentioned environmental impact assessment (EIA) prepared in accordance with local or national regulations, environmental impact assessments of wastewater treatment plants are also carried out as part of various types of expert investigations e.g., [72] and research. This type of work is observed in numerous scientific articles, for example, [12,73,74,104 or 105]. Their scope, structure and level of detail are determined by the authors of the publications and often result from current problems occurring in the facilities. In articles on the assessment of the impact of wastewater treatment plants on the environment, the LCA technique is often the research method—used alone [104,105] or in combination with other methods—such as in [74] with material flow analysis (MFA) and energy analysis (EA). For example, the work [74] analyzed the environmental impact of energy and chemical consumption at wastewater treatment plants in Oslo, Norway. According to this study, global warming and acidification were the dominant effects from the chemicals and energy, respectively. In turn, the work [105] discusses the environmental assessment of wastewater treatment plants for Al-Hilla City in Iraq with the use of LCA. The greatest impact was recorded in terms of global warming, the influence of inorganic substances on the respiratory system and the consumption of non-renewable energy sources. Other methods of environmental impact assessment are also used, such as in [73], where the Fine–Kinney method was used to compare three versions of a technological system for treating wastewater from meat-processing plants. Some of the assessments of wastewater treatment plants in scientific articles are carried out in the context of energy consumption and its optimization. This is related to the ongoing climate change and the attempts to minimize the amounts of greenhouse gases produced. The articles [106,107] are examples here. The work [107] shows a multi-step methodology for the appraisal of the energetic aspects of wastewater treatment, introduced at a wastewater treatment plant in Italy. In turn, the article [106] proposes a multi-step simulation-based methodology for fully relating treatment processes to energy demand and energy production. This work also assumes a scenario-based optimization approach and applies it to the same facility (the WWTP in Castiglione Torinese, Italy). There are also assessments that take into account only one type of environmental impact—e.g., noise in [72] or odors [108]. The work [108] shows the findings of the appraisal of the odor intensity (conducted via sensory studies according to a six-stage scale) and the measurement of odor concentration (with the use of portable field olfactometers) after the finishing of a modernization project, and compares them with similar research carried out before the start of this investment. After modernization, a meaningful diminution in the concentration of the odor emitted from the sludge dewatering building, sludge containers and sewage channel was determined [108].

## 5. Conclusions

A wastewater treatment plant has an environmental impact during both construction and operation. Therefore, there is a need to assess this impact not only at the design and construction stage, but also at the facility operation stage. While such assessments are frequent at the investment planning stage, they tend to be neglected in the operational phase. On the other hand, control activities are carried out in the context of compliance with certain regulations concerning, for example, emissions.

Various methodologies are available for assessing the environmental impacts of wastewater treatment plants. Their scopes and degrees of detail vary. Although some of them are applicable to both designed and existing facilities (e.g., EIA, LCA, and green building certification systems), their use is often limited to the planning and design stage. The reason for this can be associated with two main factors—the obligation of assessment for planned facilities and the difficulty of meeting individual green building requirements for existing buildings. Here, the environmental management systems for existing facilities and functioning companies managing wastewater treatment plants seem to be a solution. These systems are voluntary, while the incentive for owners is the prestige from their implementation and the savings (by reducing the use of energy, raw materials and water) that go hand in hand with minimizing the impact on the environment. Moreover, these systems make it possible to organize and systematize the introduction of innovations in wastewater treatment plants and stimulate the constant search for new solutions and improvements. An interesting solution is LCA analysis, which is gradually gaining more and more interest. It is used both for the assessment and comparison of various technological cycles for planned wastewater treatment plants, and for the evaluation and comparison of existing facilities. The need to purchase software may be a problem, but there are also free programs. On the other hand, the assessments carried out by scientists are relatively rare cases, with scopes adapted to the current needs of the wastewater treatment plants and the problems to be solved.

Conducting environmental impact assessments of WWTPs during their operation seems to be a good course of action for the future. It is important to treat this issue as a whole, i.e., to take into account, apart from the typical impacts of WWTPs on environmental components, the impacts generated by the operation of their buildings. A problem may be the lack of sufficient incentives and motivation to carry out such environmental assessments once the facility is in place. The introduction of obligatory inspections of WWTPs covering all environmental aspects may be considered, but it seems much more appropriate to lead activating activities for the voluntary assessment of these facilities. It is from such cases that one can expect breakthrough solutions that will also inspire other decision-makers.

It should be emphasized that each activity aimed at assessing the impact of the operated wastewater treatment plant, regardless of the scale and scope, has a potential environmental benefit. Moreover, the analysis of the available environmental impact assessments carried out by WWTPs can provide an impulse and invitation to modernize other existing treatment plants and introduce technical and technological innovations in these facilities, all for the benefit of the environment. It should be remembered that the best available techniques (BAT) used at the time of designing the treatment plant may sometimes become obsolete after years of operation of the facility.

The suggested direction of development for the assessment of the environmental impact of wastewater treatment plants is the improvement of the LCA technique, and the implementation of environmental management systems, with the supplementation of these actions through green building certification. Only a holistic approach to the issue will enable all the environmental aspects to be taken into account and thus contribute to the maximization of the subsequent environmental benefits. It is also advisable, if possible, to take steps to develop and implement a unified method of assessing the impact of sewage treatment plants on the environment. At the same time, activities that encourage and stimulate the uptake of such challenges should be introduced. By taking such steps, in the future, the processes carried out in wastewater treatment plants will become even more "clean", and the treatment plants themselves will become more environmentally sustainable facilities. In this aspect, the introduction of circular economy solutions in wastewater treatment plants is also of great importance.

**Author Contributions:** J.B.—conceptualization, methodology, formal and critical data analysis, and article writing and editing; K.B.—substantive consultations, data curation and article editing; M.T.—substantive consultations and article writing and editing. All authors have read and agreed to the published version of the manuscript.

**Funding:** This research received no external funding.

**Acknowledgments:** The authors would like to thank the anonymous reviewers for their helpful comments.

**Conflicts of Interest:** The authors declare no conflict of interest.

## References


1. *Act of 3 October 2008 On the Provision of Information On the Environment and Its Protection, the Public Participation in Environmental Protection and On Environmental Impact Assessment*; Consolidated Text; Dziennik Ustaw, Rządowe Centrum Legislacji: Warsaw, Poland, 2008; No 199.1227. (In Polish). Available online: http://isap.sejm.gov.pl/isap.nsf/DocDetails.xsp?id=WDU20081991227 (accessed on 15 December 2020).
2. *Regulation of the Council of Ministers of September 10, 2019 On Projects That May Significantly Affect the Environment*; Dziennik Ustaw, Rządowe Centrum Legislacji: Warsaw, Poland, 2019; p. 1839, (In Polish). Available online: https://isap.sejm.gov.pl/isap.nsf/download.xsp/WDU20190001839/O/D20191839.pdf (accessed on 19 October 2020).
3. *EN ISO 14001:2015 Environmental Management Systems. Requirements with Guidance for Use*; International Organization for Standardization: Geneva, Switzerland, 2015.
4. Tabor, A. (Ed.) *Staszczyszyn B. Basic issues of Environmental Management in Environmental Management Systems-ISO 14000/EMAS. Quality Concept, eco-Management and Audit System EMAS, Environmental Management-Basic Issues, Implementation, Maintenance and Improvement*; Center for Training and Organization of Quality Systems of the Tadeusz Kościuszko Cracow University of Technology: Kraków, Poland, 2009; Volume 3, pp. 53–76. (In Polish)
5. The Act of July 20, 2017. In *Water Law Act*; Consolidated Text; Dziennik Ustaw, Rządowe Centrum Legislacji: Warsaw, Poland, 2017; p. 1566, (In Polish). Available online: https://isap.sejm.gov.pl/isap.nsf/DocDetails.xsp?id=WDU20170001566 (accessed on 15 December 2020).
6. *Regulation of the Minister of the Environment of 27 August 2014 on Types of Installations that May Cause Significant Pollution of Individual Natural Elements or the Environment*; Dziennik Ustaw, Rządowe Centrum Legislacji: Warsaw, Poland, 2014; p. 1169, (In Polish). Available online: http://isap.sejm.gov.pl/isap.nsf/download.xsp/WDU20140001169/O/D20141169.pdf (accessed on 15 October 2020).
7. Central Statistical Office. Available online: https://bdl.stat.gov.pl (accessed on 19 November 2020).
8. Regulation (EC) No 1221/2009 of the European Parliament and of the Council of 25 November 2009 on the Voluntary Participation by Organizations in a Community Eco-Management and Audit Scheme (EMAS), Repealing Regulation (EC) No 761/2001 and Commission Decisions 2001/681/EC and 2006/193/EC; Consolidated Version: 9.01.2019. *Off. J. Eur. Union* **2009**, 1. Available online: https://eur-lex.europa.eu/legal-content/EN/TXT/?uri=CELEX%3A02009R1221-20190109 (accessed on 19 October 2020).
9. Barbusinski, K.; Kalemba, K.; Kasperczyk, D.; Urbaniec, K.; Kozik, V. Biological methods for odor treatment—A review. *J. Clean. Prod.* **2017**, *52*, 223–241. [CrossRef]
10. Bak, A.; Kozik, V.; Dybal, P.; Sulowicz, S.; Kasperczyk, D.; Kus, S.; Barbusinski, K. Abatement robustness of volatile organic compounds using compact trickle-bed bioreactor: Biotreatment of styrene, ethanol and dimethyl sulfide mixture in contaminated airstream. *Int. Biodeterior. Biodegrad.* **2017**, *119*, 316–328. [CrossRef]
11. Michałkiewicz, M. Wastewater and their negative role in the environment. *Water Technol.* **2018**, *5*, 30–33.
12. Adamus-Białek, W.; Wawszczak, M.; Świercz, A. Impact of sewage treatment plant on local environment. *Proc. ECOpole* **2015**, *9*, 397–404.
13. Carducci, A.; Tozzi, E.; Rubulotta, E.; Casini, B.; Cantiani, L.; Rovini, E.; Muscillo, M.; Pacini, R. Assessing airborne biological hazard from urban wastewater treatment. *Water Res.* **2000**, *34*, 1173–1178. [CrossRef]
14. Cyprowski, M.; Krajewski, J.A. Harmful agents in municipal wastewater treatment plants. *Med. Pr.* **2003**, *54*, 73–80.
15. Sadowski, J. Minimizing environmental noise on the example of the sewage treatment plant in Sztum. *Inż. I Apar. Chem.* **2012**, *51*, 251–253.
16. Skorupski, W. *Protection of Waters Against Pollution, Occupational Health and Safety in Water and Sewage Management*; Central Center of Staff Training for Environmental Protection and Water Management: Dębe, Poland, 1990; pp. 1–32.
17. Fighir, D.; Teodosiu, C.; Fiore, S. Environmental and energy assessment of municipal wastewater treatment plants in Italy and Romania: A comparative study. *Water* **2019**, *11*, 1611. [CrossRef]
18. Urban, S.; Dzikuć, M. The impact of electricity production in coal-fired power plants on environment. *Ekon. I Śr.* **2013**, *2*, 84–92.
19. *PGE Górnictwo i Energetyka Konwencjonalna S.A, Environmental declaration according to the EMAS*, XII ed.; Oddział Elektrownia: Opole, Poland, 2019; pp. 1–43. Available online: http://emas.gdos.gov.pl/files/artykuly/24009/1_2019_DS_PGE-Opole_icon.pdf (accessed on 19 October 2020).
20. Law, Y.; Ye, L.; Pan, Y.; Yuan, Z. Nitrous oxide emissions from wastewater treatment processes. *Philos. Trans. R. Soc. Lond. B Biol. Sci.* **2012**, *367*, 1265–1277. [CrossRef] [PubMed]
21. Mikosz, J. Analysis of greenhouse gas emissions and the energy balance in a model municipal wastewater. *Desalin. Water Treat.* **2016**, *57*, 28551–28559. [CrossRef]
22. Foley, J.; De Haas, D.; Hartley, K.; Lant, P. Comprehensive life cycle inventories of alternative wastewater treatment systems. *Water Res.* **2010**, *44*, 1654–1666. [CrossRef] [PubMed]
23. Thomas, M.; Białecka, B.; Zdebik, D. Removal of organic compounds from wastewater originating from the production of printed circuit boards by UV-Fenton method. *Arch. Environ. Prot.* **2017**, *43*, 39–49. [CrossRef]
24. Thomas, M.; Zdebik, D.; Białecka, B. Treatment of wastewater from the photochemical production of printed circuit boards by using Fenton reagent after addition of calcium peroxide. *Przem. Chem.* **2016**, *1*, 134–139.
25. Thomas, M.; Zdebik, D.; Białecka, B. Treatment of wastewater from the photochemical production of printed circuit boards by using Fenton reagent after addition of disodium percarbonate. *Przem. Chem.* **2015**, *94*, 924–929.

26. Nabi, G.R.; Torabian, A.; Ehsani, H.; Razmkhah, N. Evaluation of industrial dyeing wastewater treatment with coagulants and polyelectrolyte as a coagulant aid. *Iran. J. Environ. Health. Sci. Eng.* **2007**, *4*, 29–36.
27. Alexander, J.T.; Hai, F.I.; Al-aboud, T.M. Chemical coagulation-based processes for trace organic contaminant removal: Current state and future potential. *J. Environ. Manag.* **2012**, *111*, 195–207. [CrossRef]
28. Thomas, M.; Zdebik, D.; Białecka, B. Using sodium trithiocarbonate to precipitate heavy metals from industrial wastewater-from the laboratory to industrial scale. *Pol. J. Environ. Stud.* **2018**, *27*, 1–11. [CrossRef]
29. Thomas, M.; Zdebik, D.; Białecka, B. Use of sodium trithiocarbonate for remove of chelated copper ions from industrial wastewater originating from the electroless copper plating process. *Arch. Environ. Prot.* **2018**, *44*, 32–42.
30. Thomas, M.; Zdebik, D.; Białecka, B. Removal of copper, nickel and tin from model and real industrial wastewater using sodium trithiocarbonate. The negative impact of complexing compounds. *Arch. Environ. Prot.* **2018**, *44*, 33–47.
31. Barbusiński, K.; Parzentna-Gabor, A.; Kasperczyk, D. Removal of odors (mainly $H_2S$ and $NH_3$) using biological treatment methods. *Clean. Technol.* **2020**, *2*.
32. Byliński, H.; Gębicki, J.; Namieśnik, J. Evaluation of Health Hazard Due to Emission of Volatile Organic Compounds from Various Processing Units of Wastewater Treatment Plant. *Int. J. Environ. Res. Public Health* **2019**, *16*, 1712. [CrossRef] [PubMed]
33. Kobyłko, G. (Ed.) *Pro-Ecological Enterprise Management*; Publishing House of the Oskar Lange University of Economics in Wrocław: Wrocław, Poland, 2000.
34. Zębek, E. Environmental impact assessment and water protection against pollution. *Stud. Prawnoustr.* **2012**, *18*, 173–183.
35. *Council Directive 85/337/EEC of 27 June 1985 on the Assessment of the Effects of Certain Public and Private Projects on the Environment*; Consolidated Version: 25.6.2009; Official Journal of the European Union, 5 July 1985; Volume 175, p. 40. Available online: https://eur-lex.europa.eu/legal-content/EN/TXT/?uri=CELEX%3A01985L0337-20090625 (accessed on 15 December 2020).
36. *Directive 2011/92/EU of the European Parliament and of the Council of 13 December 2011 on the Assessment of the Effects of Certain Public and Private Projects on the Environment (Text with EEA Relevance)*; Consolidated Version: 15.5.2014; Official Journal of the European Union, 28 January 2012; Volume 26, p. 1. Available online: https://eur-lex.europa.eu/legal-content/EN/TXT/?uri=CELEX%3A02011L0092-20140515 (accessed on 15 December 2020).
37. *Council Directive 92/43/EEC of 21 May 1992 on the Conservation of Natural Habitats and of Wild Fauna and Flora*; Consolidated Version 1.7.2013; Official Journal of the European Union, 22 July 1992; Volume 206, p. 7. Available online: https://eur-lex.europa.eu/legal-content/EN/TXT/?uri=CELEX%3A01992L0043-20130701 (accessed on 15 December 2020).
38. *Directive 2009/147/EC of the European Parliament and of the Council of 30 November 2009 on the Conservation of Wild Birds*; Consolidated Version: 26.6.2019; Official Journal of the European Union, 26 January 2010; Volume 20, p. 7. Available online: https://eur-lex.europa.eu/legal-content/EN/TXT/?uri=CELEX%3A02009L0147-20190626 (accessed on 15 December 2020).
39. *Directive 2008/50/EC of the European Parliament and of the Council of 21 May 2008 on Ambient Air Quality and Cleaner Air for Europe*; Consolidated Version: 18.9.2015; Official Journal of the European Union, 11 June 2008; Volume 152, p. 1. Available online: https://eur-lex.europa.eu/legal-content/EN/TXT/?uri=CELEX%3A02008L0050-20150918 (accessed on 15 December 2020).
40. *Directive 2000/60/EC of the European Parliament and of the Council of 23 October 2000 Establishing a Framework for Community Action in the Field of Water Policy*; Consolidated Version: 20.11.2014; Official Journal of the European Union, 22 December 2000; Volume 327, p. 1. Available online: https://eur-lex.europa.eu/legal-content/EN/TXT/?uri=CELEX%3A02000L0060-20141120 (accessed on 15 December 2020).
41. *Directive 2008/105/EC of the European Parliament and of the Council of 16 December 2008 on Environmental Quality Standards in the Field of Water Policy, Amending and Subsequently Repealing Council Directives 82/176/EEC, 83/513/EEC, 84/156/EEC, 84/491/EEC, 86/280/EEC and Amending Directive 2000/60/EC of the European Parliament and of the Council*; Consolidated Version: 13.9.2013; Official Journal of the European Union, 24 December 2008; Volume 348, p. 84. Available online: https://eur-lex.europa.eu/legal-content/EN/TXT/?uri=CELEX%3A02008L0105-20130913 (accessed on 15 December 2020).
42. *Directive 2006/118/EC of the European Parliament and of the Council of 12 December 2006 on the Protection of Groundwater against Pollution and Deterioration*; Consolidated Version: 11.7.2014; Official Journal of the European Union, 27 December 2006; Volume 372, p. 19. Available online: https://eur-lex.europa.eu/legal-content/EN/TXT/?uri=CELEX%3A02006L0118-20140711 (accessed on 15 December 2020).
43. *Directive 2010/75/EU of the European Parliament and of the Council of 24 November 2010 on Industrial Emissions (Integrated Pollution Prevention and Control) (Recast), Text with EEA Relevance*; Consolidated Version: 6.1.2011; Official Journal of the European Union, 17 December 2010; Volume 334, p. 17. Available online: https://eur-lex.europa.eu/legal-content/EN/TXT/?uri=CELEX%3A02010L0075-20110106 (accessed on 15 December 2020).
44. *Directive 2008/98/EC of the European Parliament and of the Council of 19 November 2008 on Waste and Repealing Certain Directives, Text with EEA Relevance*; Consolidated Version: 5.7.2018; Official Journal of the European Union, 22 November 2008; Volume 312, p. 3. Available online: https://eur-lex.europa.eu/legal-content/EN/TXT/?uri=CELEX%3A02008L0098-20180705 (accessed on 15 December 2020).
45. *Directive 2012/18/EU of the European Parliament and of the Council of 4 July 2012 on the Control of Major-Accident Hazards Involving Dangerous Substances, Amending and Subsequently Repealing Council Directive 96/82/EC, Text with EEA Relevance*; Official Journal of the European Union, 24 July 2012; Volume 197, pp. 1–37. Available online: https://eur-lex.europa.eu/legal-content/EN/TXT/?uri=CELEX%3A32012L0018 (accessed on 15 December 2020).

46. *Council Directive of 21 May 1991 Concerning Urban Waste Water Treatment (91/271/EEC)*; Consolidated Version: 1.1.2014; Official Journal of the European Union, 30 May 1991; Volume 135, p. 40. Available online: https://eur-lex.europa.eu/legal-content/EN/TXT/?uri=CELEX%3A01991L0271-20140101 (accessed on 15 December 2020).
47. *Council Directive of 12 June 1986 on the Protection of the Environment, and in Particular of the Soil, when Sewage Sludge is Used in Agriculture (86/278/EEC)*; Official Journal of the European Union, 4 July 1986; Volume 181, p. 6, Consolidated Version: 4.7.2018; Available online: https://eur-lex.europa.eu/legal-content/EN/TXT/?uri=CELEX%3A01986L0278-20180704 (accessed on 15 December 2020).
48. *Directive 2004/35/CE of the European Parliament and of the Council of 21 April 2004 on Environmental Liability with Regard to the Prevention and Remedying of Environmental Damage (Consolidated Version)*; Consolidated Version: 26.6.2019; Official Journal of the European Union, 30 April 2004; Volume 143, p. 56. Available online: https://eur-lex.europa.eu/legal-content/EN/TXT/?uri=CELEX%3A02004L0035-20190626 (accessed on 15 December 2020).
49. Available online: https://ec.europa.eu/environment/soil/index_en.htm (accessed on 19 December 2020).
50. Butler, S. *Environmental Assessment Wastewater System Improvements, City of Sterling, Logan County, Colorado, ERO Resources Corporation, Denver, Colorado, USA, April 2020*; pp. 1–42. Available online: https://cms6.revize.com/revize/sterlingco/Sterling%20WW%20System%20Improvements%20%20Environmental%20Assessment.pdf (accessed on 19 October 2020).
51. Solutions Trak. Available online: https://www.solutionstrak.com/blog/environmental-aspects-impacts/ (accessed on 19 October 2020).
52. Croner-i. Available online: https://app.croneri.co.uk/topics/environmental-aspects-and-impacts/indepth (accessed on 19 October 2020).
53. Marcinkowski, A.; Huryń, B.; Ostrowski, K.M.; Szydłowski, M. *Eco-Management in the Enterprise*; Environmental Information Center: Warsaw, Poland, 2010; pp. 1–99. Available online: https://emas.gdos.gov.pl/files/artykuly/4947/5_Ekozarzadzanie_w_przedsiebiorstwie_podrecznik.pdf (accessed on 19 October 2020).
54. Królas, L. Experience in Establishing, Implementing, Maintaining and Improving Environmental Management Systems. In *Quality Concept, Eco-Management and Audit System EMAS, Environmental Management-Basic issues, Implementation, Maintenance and Improvement*; Tabor, A., Ed.; Center for Training and Organization of Quality Systems of the Tadeusz Kościuszko Cracow University of Technology: Kraków, Poland, 2009; pp. 125–158.
55. *ISO 14040:2006 Environmental Management-Life Cycle Assessment-Principles and Framework*; International Organization for Standardization: Geneva, Switzerland, 2006.
56. *ISO 14044:2006 Environmental Management-Life Cycle Assessment-Requirements and Guidelines*; International Organization for Standardization: Geneva, Switzerland, 2006.
57. Henclik, A.; Bajdur, W.M. The use of selected methods for assessing the impact of the life cycle of the flocculant production process synthesized from phenol-formaldehyde resin waste on the quality of the environment. *Rocz. Ochr. Srodowiska* **2011**, *13*, 1809–1822.
58. Acero, A.P.; Rodriguez, C.; Ciroth, A. *LCIA Methods. Impact Assessment Methods in Life Cycle Assessment and Their Impact Categories*; Greendelta, 2016; pp. 1–23. Available online: https://www.openlca.org/wp-content/uploads/2015/11/LCIA-METHODS-v.1.5.4.pdf (accessed on 19 October 2020).
59. Solid Works. Sustainability. Available online: https://www.solidworks.com/sustainability/sustainable-design-guide/appendix-c-the-hannover-principles.htm (accessed on 19 October 2020).
60. Cooper, J.S.; Fava, J. Life cycle assessment practitioner survey: Summary of results. *J. Ind. Ecol.* **2006**, *10*, 12–14. [CrossRef]
61. Open LCA. Available online: http://www.openlca.org/contributors/ (accessed on 19 October 2020).
62. CMLCA. Available online: http://www.cmlca.eu/ (accessed on 19 October 2020).
63. Goedkoop, M.; Spriensma, R. *The Eco-Indicator 99. A Damage Oriented Method for Life Cycle Impact Assessment. Methodology Annex*; PRé Consultants B.V., 2001; pp. 1–83. Available online: www.pre.nl (accessed on 19 October 2020).
64. Collective Work; Firląg, S. (Eds.) *Sustainable Office Buildings. Projects. Legal Conditions*; Technological Solutions; PWN PLGBC: Warsaw, Poland, 2018; pp. 1–404.
65. USGBC. Available online: https://www.usgbc.org/leed/why-leed (accessed on 19 October 2020).
66. U.S. Green Building Council, LEED v4 for Building Design and Construction, Updated 2019. Available online: https://www.usgbc.org/sites/default/files/LEED%20v4%20BDC_07.25.19_tracked.pdf (accessed on 19 October 2020).
67. U.S. Green Building Council, LEED v4.1 for Building Design and Construction, July 2020. Available online: https://www.usgbc.org/leed/v41#bdc (accessed on 19 October 2020).
68. USGB. Available online: https://www.usgbc.org/leed (accessed on 18 January 2021).
69. BREEAM UK. New Constructions. Available online: https://www.breeam.com (accessed on 19 October 2020).
70. Bąk, J.; Radziszewska–Zielina, E.; Zielina, M. Analysis of the spatial structure of green building in the aspect of selected environmental issues on the example of the city of Cracow (Poland). In *Proceeding of the IOP Conference Series: Material Science and Engineering*; 2019; Available online: https://iopscience.iop.org/article/10.1088/1757-899X/603/4/042099 (accessed on 19 October 2020).
71. BREEAM UK New Construction. Non-domestic buildings (United Kingdom) Technical Manual SD5078: BREEAM UK New Construction 2018 1.0. Available online: http://www.breeam.com/NC2018/content/resources/output/10_pdf/a4_pdf/print/nc_uk_a4_print_mono/nc_uk_a4_print_mono.pdf (accessed on 19 October 2020).

72. Babicz, W.; Jaruszewska, K. *Measurement and Computational Expertise in the Identification of Noise Sources in the Sewage Treatment Plant in Lublin*; KFB Polska sp. z o.o.: Wrocław, Poland, 2011; pp. 1–34. Available online: http://www.mpwik.lublin.pl/imgs_upload2/przetargi/2012/SEKTOROWE/69_Ograniczenie_emisji_halasu/Zal_5_Ekspertyza%20pomiarowo_obliczeniowa.pdf (accessed on 19 October 2020).
73. Yapıcıoğlu, P.S. Environmental impact assessment for a meat processing industry in Turkey: Wastewater treatment plant. *Water Pract. Technol.* **2018**, *13*, 692–704. [CrossRef]
74. Venkatesh, G.; Brattebø, H. Environmental impact analysis of chemicals and energy consumption in wastewater treatment plants: Case study of Oslo, Norway. *Water Sci. Technol.* **2011**, *63*, 1018–1031. [CrossRef]
75. Collective Work. *Environemntal Impact Assessment: Sewage Treatment Plant for Port Said, UNEP Regional Seas Reports and Studies No. 133*; UNEP, 1991; pp. 1–32. Available online: https://wedocs.unep.org/handle/20.500.11822/7606 (accessed on 19 October 2020).
76. RAMBOLL. Minsk Wastewater Treatment Plant Environmental and Social Impact Assessment. April 2017, pp. 1–439. Available online: https://www.ebrd.com/documents/environment/esia-49483-supp.pdf (accessed on 19 October 2020).
77. Ratajczak, M. *Environmental Impact Report for the Project: Construction of a Sewage Treatment Plant in Bródki*; Commune Czerwiensk: Enviposse, Poland, 2017; pp. 1–105. Available online: czerwiensk.rios.pl (accessed on 19 October 2020).
78. Sroka, P. *Environmental Impact Assessment Report for the Project: Construction of the Sewage Treatment Plant in Józefów*; Envi-CONSULT Biuro Analiz Środowisowych: Zagórze, Poland, 2013; pp. 1–114. Available online: http://jozefow.bip.eur.pl/public/get_file_contents.php?id=262856 (accessed on 19 October 2020).
79. Collective Work, Report on the Environmental Impact of the Investment: Modernization and Extension of the Czajka Sewage Treatment Plant, Proeko Sp. z o.o., Energoprojekt-Warszawa S.A., Poland. 2007, pp. 1–30. Available online: https://www.eib.org/attachments/pipeline/20070356_nts_pl.pdf (accessed on 19 October 2020).
80. Smuszkiewicz, K.; Gąsak, I.; Bednarz, M.; Łakomski, T.; Gierżatowicz, R. *Environmental Impact Report for a Project Involving the Reconstruction and Extension of the Sewage Treatment Plant in Łańcut*; EKO-GEO: Lublin, Poland, 2009; pp. 1–128. Available online: http://www.lzk-lancut.pl/przetarg/Raport%20%20o%20oddzia%C5%82ywaniu%20na%20%C5%9Brodowisko/Raport%20OO%C5%9A%20Oczyszczalnia%20%C5%81a%C5%84cut.pdf (accessed on 19 October 2020).
81. Nytko, K.; Bartkowska, I. Analysis of environmental impact reports of municipal wastewater treatment plants. *Ekon. Śr.* **2013**, *2*, 264–273.
82. Fuentes-Bargues, J.L. Review of the environmental impact assessment process of wastewater treatment plants in Spain. *Environ. Prot. Eng.* **2018**, *44*, 23–41.
83. Jajac, N.; Marović, I.; Rogulj, K.; Kilić, J. Decision support concept to selection of wastewater treatment plant location-the case study of town of Kutina, Croatia. *Water* **2019**, *11*, 717. [CrossRef]
84. Pasqualino, J.C.; Meneses, M.; Abella, M.; Castells, F. LCA as a decision support tool for the environmental improvement of the operation of a municipal wastewater treatment plant. *Environ. Sci. Technol.* **2009**, *43*, 3300–3307. [CrossRef] [PubMed]
85. Amaral, K.; Aisse, M.; Possetti, G. Sustainability assessment of sludge and biogas management in wastewater treatment plants using the LCA technique. Rev. Ambient. Água. *An Interdiscip. J. Appl. Sci.* **2019**, *14*, 1–14. [CrossRef]
86. Zhao, C.H.; Zhang, Y.; Liang, K.; Li, J. Environmental impact analysis of wastewater treatment process based on life cycle assessment. In Proceedings of the International Conference on Education, Management and Computing Technology (ICEMCT 2015), Tianjin, China, 13–14 June 2015; pp. 1614–1617. [CrossRef]
87. Kamble, S.; Singh, A.; Kazmi, A.; Starkl, M. Environmental and economic performance evaluation of municipal wastewater treatment plants in India: A life cycle approach. *Water Sci. Technol.* **2019**, *79*, 1102–1112. [CrossRef] [PubMed]
88. Buonocore, E.; Mellino, S.; De Angelis, G.; Liu, G.; Ulgiati, S. Life cycle assessment indicators of urban wastewater and sewage sludge treatment. *Ecol. Indic.* **2018**, *94*, 13–23. [CrossRef]
89. Bajdur, W.; Henclik, A.; Skowron-Grabowska, B.; Iwaszczuk, N. LCA application in the assessment of new technologies of industrial effluents treatment. *Desalin. Water Treat.* **2015**, *57*, 1–9. [CrossRef]
90. Bajdur, W.; Henclik, A.; Ščurek, R.; Sikorová, K. Ecological life cycle assessment of modified novolaks waste used in industrial wastewater treatment. *Trans. VŠB-Tech. Univ.* **2017**, *12*, 41–50. [CrossRef]
91. Godin, D.; Bouchard, C.; Vanrolleghem, P.A. Net environmental benefit: Introducing a new LCA approach on wastewater treatment systems. *Water Sci. Technol.* **2012**, *65*, 1624–1631. [CrossRef]
92. CHSPWIK. Available online: https://www.chspwik.pl/o-firmie/iso-1400 (accessed on 19 October 2020).
93. MWIO. Available online: https://www.mwio.pl/main/jakosc_srodowisko_ekologia.html (accessed on 19 October 2020).
94. WOBI. Available online: https://www.wobi.pl/o-firmie/zintegrowany-system-zarzadzania.html (accessed on 19 October 2020).
95. Waterworks in Kraków. Available online: https://wodociagi.krakow.pl/admin/files/Files/woda_i_my/2019/WODA_89_opt.pdf (accessed on 19 October 2020).
96. EMAS. Available online: https://emas.gdos.gov.pl/lista-rejestru-emas (accessed on 19 October 2020).
97. *Environmental Declaration of Przedsiębiorstwo Wodociągów i Kanalizacji Okręg Częstochowski S.A. in Częstochowa*, 4th ed.; Przedsiębiorstwo Wodociągów i Kanalizacji Okręgu Częstochowskiego SA w Częstochowie: Częstochowa, Poland, 2019; pp. 1–68. Available online: https://www.pwik.czest.pl/20193 (accessed on 19 October 2020).
98. GreenBookLive. Available online: https://www.greenbooklive.com (accessed on 20 October 2020).
99. European Investment Bank. Available online: https://www.eib.org/en/stories/once-upon-a-time# (accessed on 20 October 2020).
100. USGBC. Available online: https://www.usgbc.org/projects (accessed on 20 October 2020).

101. Canada Green Building Council. Available online: https://leed.cagbc.org/LEED/projectprofile_EN.aspx (accessed on 20 October 2020).
102. Stantec. Available online: https://www.stantec.com/en/news/2017/dryden-stantec-designed-wwtp-leed-silver (accessed on 20 October 2020).
103. Lund, S. Sustainable Through and Through, Treatment Plant Operator Magazine, September 2017. Available online: https://www.tpomag.com/editorial/2017/09/entire_clean-water_plant_earns_leed_recognition (accessed on 20 October 2020).
104. Al-Dosary, S.; Galal, M.M.; Abdel-Halim, H. Environmental impact assessment of wastewater treatment plant (Zenien and 6th of October WWTP). *Int. J. Curr. Microbiol. Appl. Sci.* **2015**, *4*, 953–964.
105. Alanbari, M.A.; Alazzawi, H.Q.; Al-Ansari, N.A.; Knutsson, S. Environmental Assessment of Al-Hilla City Wastewater Treatment Plants. *J. Civ. Eng. Arch.* **2015**, *9*, 749–755.
106. Borzooei, S.; Campo, G.; Cerutti, A.; Meucci, L.; Panepinto, D.; Ravina, M.; Riggio, V.; Ruffino, B.; Scibilia, G.; Zanetti, M.C. Optimization of the wastewater treatment plant: From energy saving to environemntal impact mitigation. *Sci. Total Environ.* **2019**, *691*, 1182–1189. [CrossRef] [PubMed]
107. Panepinto, D.; Fiore, S.; Zappone, M.; Genon, G.; Meucci, L. Evaluation of the energy efficiency of a large wastewater treatment plant in Italy. *Appl. Energy* **2016**, *161*, 404–411.
108. Kulig, A.; Szyłak-Szydłowski, M. Assessment of the effects of wastewater treatment plant modernization by means of the field olfactometry method. *Water* **2019**, *11*, 2367. [CrossRef]

MDPI
St. Alban-Anlage 66
4052 Basel
Switzerland
Tel. +41 61 683 77 34
Fax +41 61 302 89 18
www.mdpi.com

*Clean Technologies* Editorial Office
E-mail: cleantechnol@mdpi.com
www.mdpi.com/journal/cleantechnol

www.ingramcontent.com/pod-product-compliance
Lightning Source LLC
Chambersburg PA
CBHW062105220926
51850CB00010B/2896
* 9 7 8 3 0 3 6 5 3 4 8 1 7 *